BACTERIOLOGY RESEARCH DEVELOPMENTS

LISTERIA MONOCYTOGENES

INCIDENCE, GROWTH BEHAVIOR AND CONTROL

BACTERIOLOGY RESEARCH DEVELOPMENTS

Additional books in this series can be found on Nova's website under the Series tab.

Additional e-books in this series can be found on Nova's website under the e-book tab.

BACTERIOLOGY RESEARCH DEVELOPMENTS

LISTERIA MONOCYTOGENES

INCIDENCE, GROWTH BEHAVIOR AND CONTROL

TIFFANY VICARIO
EDITOR

NOVA BIOMEDICAL

New York

Library of Congress Cataloging-in-Publication Data

Library of Congress Control Number: 2015949939

ISBN: 978-1-63483-765-1

Published by Nova Science Publishers, Inc. † New York

Contents

PREFACE

This book provides readers with the latest developments in *Listeria monocytogenes* research. Topics covered include its growth in biofilms in the food industry; molecular mechanisms involved in the food-related subsistence of *Listeria Monocytogenes*; an epidemiological update and the control possibilities for outbreaks of *L. Monocytogenes*; how to control *L. Monocytogenes* biofilms on food contact surfaces; methods for killing activities of macrophages against *L. Monocytogenes* infection; and a review of some of the most exciting developments in the analysis of the interaction between *L. Monocytogenes* and the host GI tract.

Chapter 1 – The ubiquitous nature of *Listeria monocytogenes* has been repeatedly verified through the isolation from a wide variety of environmental sources. The specific interest that this microorganism has attracted resulted from the ability to adapt to a series of stresses that may be faced upon colonization of substrates that are considered for human nutrition. Moreover, upon ingestion, a series of genes are activated and a remarkable intracellular life cycle commences through which the extracellular environment along with the humoral effectors of the human immune system are avoided resulting in adverse health effects to the host that are collectively referred to as listeriosis. In this chapter, an update on the knowledge of the molecular mechanisms involved in the food-related subsistence of *L. monocytogenes* is offered.

Chapter 2 – *Listeria monocytogenes* is an important foodborne pathogen not because it causes large numbers of symptomatic cases but because of its relatively high case–fatality rate (20-30%). Listeriosis appears to be primarily a foodborne infection and is particularly a problem on foods that are not cooked, including ready-to-eat (RTE) meats, soft cheeses, and unpasteurized dairy products, as well as vegetables. Thermal processing of milk and meat products will destroy *L. monocytogenes* but post-processing contamination does occur. *L. monocytogenes* that contaminates foods may originate in soils from farms, in slaughtering facilities, in food processing plants and other equipments used in food manufacture. Elimination of Listeria from many environmental sources can be challenging because of the resistant, persistent biofilms formed by this microorganism. Bacterial biofilms are generally described as surface-associated bacterial communities forming microcolonies surrounded by a matrix of exopolymers. Biofilm formation can be separated into four sequential steps: (*i*) conditioning of the surface, (*ii*) initial adherence, (*iii*) physical irreversible adherence that involves the production of exopolymers that fix the cells, and (*iv*) growth of the microorganisms, which form microcolonies, and coaggregation leading to establishment of the biofilm. Such structures are dynamic systems in which cells grow, die, and/or are released, allowing contamination of other surfaces. Moreover, biofilms confer to *L. monocytogenes* a greater resistance to antimicrobial agents and they allow microorganisms to resist to environmental changes and cleaning/disinfection treatments. In the following, through an up-to-date review of (personal and non) published data, the growth of *L. monocytogenes* in biofilm in the food industry will be discussed.

Chapter 3 – *Listeria monocytogenes* is a gram positive bacterium which causes illness in animals and humans. This foodborne pathogen is the etiologic agent of listeriosis and can lead to spontaneous abortion in pregnant women, meningitis and encephalitis in newborns, children and immuno-suppressed adults, and gastroenteritis in health persons. *L. monocytogenes* is frequently detected in environmental samples and can survive for long periods in soil, water, feed and feces, providing be a source of contamination for the food and water chain CDC (Center for Disease Control and Prevention).

L. monocytogenes and *L. innocua* are frequently isolated from food processing plants. However, most of the cases of listeriosis transmitted through food are due to the contamination of raw or cooked foods with *L. monocytogenes*. The minimum infective dose of *L. monocytogenes* has not been established yet, although it has been indicated that the intake of up to 100 cells does not affect healthy consumers. The long incubation time of listeriosis makes difficult to identify the pathogen and trace the contaminated food. There are 13 serovars known of *L. monocytogenes*, and four of them (1/2a, 1/2b, 1/2c y 4b) are responsible for over 95% of reported human listeriosis cases. The major outbreaks have been caused by *L. monocytogenes* 4b. In general, mortality rate is low, however in inmuno-supressed and susceptible patients this rate is 20 – 30%.

Chapter 4 – Human listeriosis is a serious disease with severe manifestations and long term sequelae in some patients, although the relative low incidence (0.3-0.7 cases/ 100,000 population). Listeriosis data is typically restricted to developed countries, although many of these do not consider the disease as a notifiable one. European member states have reported a statistically significant increasing trend over the period 2009-2013, predominantly related with ready-to-eat foods (RTE) consumption. RTE meat-based food products (RTEMP) are one of the most consumed food products around the world. Their long shelf-lives in refrigerated storage, ease of use, as well as the fact of not requiring any heat-treatment prior to consumption, makes them an important source of food-borne disease, especially by *L. monocytogenes*. This ubiquitous bacterium, widely distributed throughout the natural environment, has a psychrotrophic character and a long persistence capability in the food processing environment due to its ability to form biofilms. These appear to be the main causes of *L. monocytogenes* high prevalence in refrigerated RTEMP. There are several ways by which *L. monocytogenes* can remain in the finished RTEMP product, namely by its recontamination following a listericidal treatment as a result of the contact with contaminated processing equipments or surfaces, in processes such as slicing, cutting, shredding and packaging. Also, this contamination might occur at several stages before consumption, namely in the manufacturing industry, retail establishments, institutional kitchens and even at household level. In retail, significant differences were found in sliced RTEMP depending on the presentation type, having the in-store packaged products a significantly higher incidence than the manufacturer packaged products. So, in this review, available data on *L. monocytogenes* in different RTEMP processing stages along the food chain will be presented. Surveys on *L. monocytogenes* in ready-to-eat foods showed that approximately 2% of RTEMP were contaminated at the end of shelf-life. But since RTEMP include a large variety of products with different technological treatments, in this review we will present data of its frequency in cooked, cured, fermented and smoked RTEMP. Preventive measures will be highlighted and discussed, namely those associated with the pathogen´s establishment and transmission from processing equipments to foods, such as accurate validation of hygiene procedures, application of "seek and destroy" method, equipment and plant design improvement, rigorous temperature control and staff training towards hygiene. Finally, we will also consider different technological strategies for *L. monocytogenes* control including: RTEMP reformulation with growth inhibitors use, technological process redesign with the adoption of emerging technologies, use of biocides in the food processing environment and antimicrobial packaging. *L. monocytogenes* control in RTEMP requires coordinated efforts along the food chain and should be regarded as part of a more vast process, in which public health policies play an important role in reducing the burden of listeriosis.

Chapter 5 – The aim of the present study was to evaluate the prevalence of *Listeria monocytogenes (L.m)* in poultry marketed in Varamin, Tehran province, Iran. A total of 584 samples, including intestinal contents of chicken (n = 370) and turkey (n = 214) were randomly collected from different wet markets, between November 2008 to July 2010. USDA method was used for isolation of *Listeria* spp and *L. m.* The *L. m.* isolates were confirmed by API *Listeria* and PCR and were seroptyped using antisera against O and H antigens. The Kirby-Bauer disc diffusion method was used to assess the antimicrobial susceptibility to 10 antibiotics (*Tetracycline, Gentamycin, Chloramphenicol, Vancomycin, Rifampin, Pen G, Kanamycin, Erythromycin, Clindamycin and Amoxicillin-clavulanic acid)* according to CLSI. Furthermore, the virulence profile of *L.m* isolates (*inl*A, *inl*B, *inl*C, *inl*J, *act*A, *hly*A, *iap*, *plc*A, *prf*A) and the presence of antimicrobial resistance genes (*tet*A, *tet*B, *van*A, *van*B, *erm*B, *cml*A, penA) was investigated by PCR. The contamination rate with *Listeria* spp in poultry was 16%. The most frequent isolated species was *L. ivanovii* (9%). *L.m* was detected in 3% of poultry samples and serovar 1/2a (61%) was the predominant among *L.m* isolates. All virulence-associated genes were detected in 85% of the isolates. The 33% of *L.m* isolates were sensitive to all tested antibiotics: the remaining *L.m* isolates were resistant to 1 to ≥ 3 antibiotics. The presence of antimicrobial resistance genes was not completely correlated with antimicrobial resistance to *Tetracycline* and *Pen.* The main results of the present study confirmed the high prevalence of *Listeria* spp. and the high rate of antimicrobial resistance of *L.m* isolates from poultry products marketed in Iran.

Chapter 6 – Currently *Listeria monocytogenes* is one of the most important food pathogens worldwide. This saprophytic microorganism is responsible for a highly fatal foodborne disease in humans and can be found inside food industries, mainly because of poor hygiene practices, unhygienic design of equipment and cross-contamination. Several studies have shown that *L. monocytogenes* is capable to attach and form biofilms on the surfaces of many materials, as metal, glass, polyethylene, and rubber, being able to persist for long periods in food plants. It has been shown that *L. monocytogenes* strains that persist in food processing environments form thicker biofilms than planktonic strains, indicating that the biofilms are an important strategy for the survival of these bacteria. For example, bacterial cells within biofilms are much more resistant to antimicrobials, detergents, and sanitizing agents than their planktonic counterparts. However, the biofilm formation and its preventive measures depends on several aspects, such as the lineage of *L. monocytogenes*, its origin of isolation, nutrient levels present in media, environmental temperature, and hydrophobicity of surfaces and cells. The purpose of this review is to study the general aspects and how to control the *L. monocytogenes* biofilms on food contact surfaces.

Chapter 7 – *Listeria monocytogenes* is a ubiquitous bacterium that causes significant foodborne disease with high mortality rates in immunecompromised adults. Recent application of transcriptomic and proteomic approaches, as well as the development of new model systems, has enabled a greater focus on factors that influence adaptation to gastrointestinal environments and adhesion to and invasion of the gastrointestinal mucosa. Here, the author reviews some of the most exciting developments in the analysis of the interaction between *L. monocytogenes* and the host GI tract.

Chapter 8 – *Listeria monocytogenes* is a facultative intracellular bacterium and hence, this bacterium is able to survive and propagate in macrophages (Mϕ). Yet, in contrast to this notion, a majority of *L. monocytogenes* is killed by Mϕ, indicating that Mϕ play a critical role in protection against *L. monocytogenes* infection. Because the number of *L. monocytogenes* engulfed and killed by Mϕ depends on whether or not Mϕ are activated, experimental listeriosis has been employed for analyzing Mϕ activities by many investigators. There are several methods for analyzing phagocytic and listericidal activities of Mϕ. Yet, in most cases expensive materials and complicated methods are required. In this review, in addition to methods recently employed for determination of phagocytic and listericidal activities of Mϕ, simple and excellent, though old-fashioned, method is also introduced.

In: Listeria monocytogenes
Editor: Tiffany Vicario

ISBN: 978-1-63483-765-1

Chapter 1

LISTERIA MONOCYTOGENES AS A FOOD CONTAMINANT: A GENOMIC PERSPECTIVE

Agni Hadjilouka, Spiros Paramithiotis[*] and Eleftherios H. Drosinos

Laboratory of Food Quality Control and Hygiene,
Department of Food Science and Human Nutrition,
Agricultural University of Athens,
Iera Odos, Athens, Greece

ABSTRACT

The ubiquitous nature of *Listeria monocytogenes* has been repeatedly verified through the isolation from a wide variety of environmental sources. The specific interest that this microorganism has attracted resulted from the ability to adapt to a series of stresses that may be faced upon colonization of substrates that are considered for human nutrition. Moreover, upon ingestion, a series of genes are activated and a remarkable intracellular life cycle commences through which the extracellular environment along with the humoral effectors of the human immune system are avoided resulting in adverse health effects to the host that are collectively referred to as listeriosis. In this chapter, an update on the knowledge of the molecular mechanisms involved in the food-related subsistence of *L. monocytogenes* is offered.

1. INTRODUCTION

Listeria monocytogenes is an ubiquitous pathogen widely distributed in soil surface, decaying vegetation, soil sewage, animal feces, river and canal waters, fertilizers, plants and animals. Being such a widespread bacterium, it contaminates a wide variety of foodstuff. Hence, human food-borne listeriosis has been linked to diverse products and surveys have revealed the presence of the pathogen in various meat products, dairy products, seafood, fruits and vegetables.

[*] Corresponding author: E-mail: sdp@aua.gr; Tel.: +30 210 5294705 Fax: +30 210 5294683

L. monocytogenes has the remarkable ability to withstand a wide variety of stresses; this ability enables it to survive and proliferate both on substrates that are considered for human nutrition as well as in the human gastrointestinal tract. This dual lifestyle has been the epicenter of intensive study. Recent advances in the field of molecular biology have enabled research to focus on the molecular mechanisms activated during both lifestyles. The majority of the studies have been performed in liquid media and only recently practical limitations have been addressed enabling studies to take place on actual food matrices. Several response strategies to environmental stimuli have been proposed that have increased our understanding of the pathogen's physiology. However, further research is nessecary in order to obtain a thorough and integrated in-depth view of all the regulatory mechanisms that enable the tuning of those physiological responses. In the next paragraphs, an update on the knowledge of the molecular mechanisms involved in the food-related subsistence of *L. monocytogenes* is offered.

2. Strain Biodiversity

L. monocytogenes strain biodiversity in foodstuff has been the epicenter of intensive study over the last decades. A wide range of techniques currently exists; the selection of which depends upon a number of factors with the type of study being the most decisive one.

Epidemiological studies, i.e., studies involving the identification of contamination sources and transmission routes requires the selection of subtyping techniques able to assist, integrate and verify the epidemiological data. In the absence of such data, misinterpretation of subtyping is very likely to occur and therefore great caution should be given. The above mentioned support is provided by a genetic marker that is accordingly selected. The genetic marker should be able to provide with adequate separation of the isolates and at the same time remain concordant with the epidemiological data; therefore it is selected taking into consideration the clonality of the microorganism under study and the period of time that is assessed. Regarding the subtyping technique selected to assess this genetic marker, it should facilitate data exchange and therefore reproducibility irrespective operator, place and time is of paramount importance (van Belkum et al. 2007). The most effective typing schemes are Pulsed-Field Gel Electrophoresis (PFGE) and Multi-Locus Variable-Number Tandem Repeat Analysis (MLVA) that are based in whole profile analysis and Multi-Locus Sequence Typing (MLST) and Clustered Regularly Interspaced Short Palindrome Repeats (CRISPRs) analysis that provide with sequence based analysis.

Biodiversity studies, on the other hand, i.e., studies aiming at the differentiation of isolates, only require high degree of reproducibility that is achieved through the careful standardization at each individual experimental step (Tyler et al. 1997). Such differentiation may be achieved by techniques such as Random Amplified Polymorphic DNA (RAPD) (Vogel et al. 2001), analysis of Amplified Fragment Length Polymorphism (AFLP) (Aarts et al. 1999; Keto-Timonen et al. 2003), restriction enzyme analysis (Giovannaci et al. 1999), enterobacterial repetitive intergenic consensus sequence analysis (Jersek et al. 1999), arbitrarily primed and infrequent restriction site analysis (Franciosa et al. 2001) and repetitive element sequence-based PCR (Zunabovic et al. 2012; Hadjilouka et al. 2014).

2.1. Epidemiological Studies

Pulsed-Field Gel Electrophoresis (PFGE)

The Pulsed-Field Gel Electrophoresis includes lysis of a standardized microbial culture suspension within agarose plugs followed by the use of restriction enzymes to cleave DNA into large fragments that are subsequently separated with gel electrophoresis in a pulsed field. It has been extensively used for the assessment of genetic diversity among various isolates from a wide range of isolation sources (Paramithiotis et al. 2008; Doulgeraki et al. 2010; 2011) as well as for subtyping of bacterial pathogens. Regarding the latter, it is distinguished by its high discriminating power and good epidemiological concordance. Therefore, an internet-based platform, namely PulseNet, has been created to enable public health laboratories to share PFGE data and thus identify outbreaks (Swaminathan et al. 2001).

The discriminatory power of PFGE may be adjusted by selecting the appropriate type and number of endonucleases. Application of a single enzyme may very often result in epidemiologically meaningless clusters or may even fail to separate unrelated isolates. Therefore the use of additional enzymes is always recommended. In the case of *L. monocytogenes* surveillance, the use of *Asc*I and secondarily of *Apa*I is required in order to effectively support epidemiological data.

PFGE is characterized by very important disadvantages; it is slow, labor-intensive, requires appropriately trained personnel and the results are often prone to misinterpretation. Regarding the latter, much effort has been made to reach a consensus. Tenover et al. (1995) provided with a glossary of terms as well as a definition of relatedness at genetic and epidemiological levels. Moreover, the importance of the clonality of the particular microorganism under study has been highlighted and the need for interpretation of the results also within the context of microorganism diversity has been documented. These recommendations were updated by Barrett et al. (2006) in an attempt to address the inconsistencies after years of application.

Multi-locus Sequencing Typing (MLST)

With multi-locus sequence typing differences between isolates are assessed in the nucleotide sequence of a number of loci. Therefore, the discriminatory power may be adjusted through the selection of the number and the type of loci used as markers.

MLST schemes have been developed for a variety of microorganisms including *Bartonella henselae* (Iredell et al. 2003), *Bordetella pertussis* (van Loo et al. 2002), *Campylobacter jejuni* (McTavish et al. 2009), *Edwardsiella tarda* (Yang et al. 2013), *Mycobacterium abscessus* (Jeon et al. 2014), *Neisserria meningitidis* (Maiden et al. 1988), *Staphylococcus aureus* (Feil et al. 2003), *Vibrio cholerae* (Kotetishvili et al. 2003), *Streptococcus suis* (King et al. 2002) and salmonellae (Kotetishvili et al. 2002; Sukhnanand et al. 2005; Topdahl et al. 2005).

In the case of *L. monocytogenes*, Salcedo et al. (2003) developed the first MLST scheme that was assessing sequences from 9 housekeeping genes, namely *abcZ*, *dat*, *ldh*, *sod*, *cat*, *dapE*, *pgm*, *bglA* and *lhkA* and was used to differentiate 62 isolates previously characterized by PFGE. Loci *sod* and *pgm* were removed from the analysis due to the low diversity observed; the remaining 7 loci were used to effectively discriminate the isolates into 29 different allelic combinations that were, in most of the cases, in concordance with PFGE analysis. A series of modifications was applied to this protocol by Ragon et al. (2008) and used for a phylogenetic assessment of *L. monocytogenes*. This modified protocol has met wide acceptance and used by Parisi et al. (2010), Mammina et al. (2013), Martin et al. (2014) and Haase et al. (2014).

A different set of loci, i.e., *actA*, *betL*, *hlyA*, *gyrB*, *pgm* and *recA* was used by Revazishvili et al. (2004) to discriminate one hundred and seventy-five isolates; it was reported that MLST succeeded in better differentiation than PFGE.

The use of virulence and virulence-associated genes instead of housekeeping ones to enhance discriminatory power was suggested by Zhang et al. (2004). More accurately, internal fragments of *prfA*, *inlB*, *inlC*, *dal*, *lisR* and *clpP* were successfully used to differentiate a set of 28 isolates. This approach was also used by Chen et al. (2005; 2007) and Cantinelli et al. (2013). Another improvement was recently presented by Miya et al. (2012). According to this new insight, tandem repeat sequence analysis of only three genomic regions was used and provided with discriminatory power comparable to MLST based on the protocol described by Zhang et al. (2004).

Multi-Locus Variable-Number Tandem Repeat Analysis (MLVA)

MLVA primarily requires the definition of the variable-number tandem-repeat sequences (VNTRs) that will be used for the comparison between the isolates. Variation in the number and the type of the VNTRs selected affects the clustering capacity and concomitantly the results obtained.

Selection of rapidly evolving loci may be useful for short-term epidemiological studies and for discrimination of closely related isolates. On the other hand, selection of slowly evolving loci may be useful in long-term epidemiological assessment and to assess genetic relatedness of isolates with different evolutionary distances. MLVA protocols have been developed and effectively applied for subtyping of many bacteria such as *Bacillus anthracis* (Keim et al. 2000), *Borderella pertussis* (Schouls et al. 2004), *Coxiella burnetti* (Svraka et al. 2006), *Enterococcus faecalis* (Titze-de-Almeida et al. 2004), *Escherichia coli* O157:H7 (Lindstedt et al. 2003b; Noller et al. 2003), *Francisella tularensis* (Farlow et al. 2001; Johansson et al. 2004), *Haemophilus influenzae* (Schouls et al. 2005), *Mycobacterium tuberculosis* (Allix et al. 2004), *M. paratuberculosis* (Overduin et al. 2004), *Neisseria meningitidis* (Yazdankhah et al. 2005), *Pseudomonas aeruginosa* (Outeniente et al. 2003), *Shigella sonnei* (Liang et al. 2007), *Yersinia pestis* (Klevytska et al. 2001) and *Salmonella* serotypes (Lindstedt et al. 2003a; Hopkins et al. 2007; Ross et al. 2009; Boxrud et al. 2007; Tien et al. 2011; Davis et al. 2009; Bergamini et al. 2011).

In Table 1 an overview of the MLVA schemes applied for *L. monocytogenes* is offered. The first protocol was developed by Murphy et al. (2007). In that study, six tandem repeats (TRs), were selected and the analysis was applied to 25 smoked salmon and 20 unrelated isolates providing with results comparable to the ones obtained by more established typing methods including PFGE.

The same protocol was also effectively applied by Dass et al. (2010) to study the ecology of *L. monocytogenes* in a cold-smoked salmon processing plant. The plant was divided into four zones and each zone was colonized by a specific type of *L. monocytogenes*. Moreover, the final product was contaminated by two types; one originating from the raw material and one colonizing the production line.

Table 1. Overview of the MLVA schemes applied for *L. monocytogenes*

name	TR sequence	Primers (5'-3')	Murphy et al. (2007)	Miya et al. (2008)	Sperry et al. (2008)	Lindstedt et al. (2008)	Chen et al. (2011)	Li et al. (2013)	Saleh-Lakha et al. (2013)
LM-TR 1	TAAAACCTA	GGCGGAAAATGGGAAGC TGCGATGGTTTGGACTGTTG	1				1		1
LM-TR 2	TATTTTTATTTAAA AATG	CCTAGAACAAATCCGCCACCAT TCGCCATTGTAAACATCCCCTATT	2						
LM-TR 3	CCGGTAGAT	GCGTGTATTAGATGCGGTTGAG GCATTCCACTATCCCCTGTTTT	3				2		
LM-TR 4	GAAGAACCAAAA	TCCGAAAAAGACGAAGAAGTAGCA TGGAACGACGGACGAAATAATAAT	4				3		2
LM-TR 5	GTAGATCCG	GTTTATGCGAATGGCGAGAT CTGGCTTCATAGGATTTACTGGAT	5						
LM-TR 6	CCAGACCCAACA	AAAAGCAGCGCCACTAACG TAAAAATCCCAATAACACTCCTGA	6					1	
TR1 (LM TR3)	CCGGTAGAT	ACATGGGAAGGGTTGCAA GGATTTACTTGATTTGACGGGT		1			4		3
TR2	CATCGG	CCATGGAAGACTACTGTTTGTA GACGGTACTGTTATCGGAAA		2			5		
TR3	TCA	GAAGGTAAAAACGGCGAAAAA ATTGCTTCTCCGTATCCCTCA		3			6		
Lm-2	TTGTAT	CGTATTGTGCGCCAGAAGTA CAGCAACGCAACAACAAACAG			1				4
Lm-3 (LM-TR1)	TAAAACCTA	CAAACCGAGATGGTGTAGCA TGGTTTTGATGGATCAACTGG			2				
Lm-8	CAGCTTTCTCAGC AG	ACGCGCAATACTATAAAGGGTGTC AGAAAAAGCGGAAGCAGATAAGAA			3				
Lm-10 (LM-TR4)	GAAGAACCAAAA	CAGATATCGATACGATTGAC CAGTTAGTATTTCCAACGTC			4			2	

Table 1. (Continued)

name	TR sequence	Primers (5'-3')	Murphy et al. (2007)	Miya et al. (2008)	Sperry et al. (2008)	Lindstedt et al. (2008)	Chen et al. (2011)	Li et al. (2013)	Saleh-Lakha et al. (2013)
Lm-11	TTGCTTGTTTTTG	GAATAAAATGCTAGATGTGG CCGATTCAAAAATAGTAAAC			5			3	5
Lm-15	CAAAAGATACAC	GGACTTAACGAATACAAAAG GCTGTTACAAGTAAAACTGG			6				
Lm-23 (TR2)	CATCGG	TATTTACGGAAAAGACGTAG CGTAACTGTCCTACCATTAG			7			4	6
Lm-32	AACACC	AAAGCTTTGCCAGTGCAAGT TTGTGACTTGGCACTTCTGG			8			5	7
LMV1	GTATTT	CGTATTGTGCGCCAGAAGTA MAMCAACRCAACAACAAACAG				1		6	
LMV2 (LM-TR5)	GTAGATCCG	TAGATGCGGTTGAGRTAGAYR CTGGMTYMATWGGATTTACTKGAT				2		7	
LMV6	AGTACCACCAACACC	AAAAGCCCCRATTGGATA CTCGCTGTTTTCTGWTTTCTTAGG				3		8	
LMV7 (Lm-3; LM-TR1)	TAAAACCTA	TCMAAAATCAAGCACAAATCACTG TAGCAAGCAWAYGCCTGTCCAKA				4		9	
LMV9	AGAAAAACC	AACGGTKRCKGATTTACTTC CTTGGYGTCGAGGCATTTA				5			
LM-Lm2V1	NR	AAGGGAGATTCCCGCTTTAG AAAAGCGGTGACACTATTTGG							8

NR: not reported.

The effectiveness of this technique was also indicated by Miya et al. (2008) that reported the development and validation of a method specifically for 4b isolates that provided with more discriminating power than an MLST scheme based on six virulence and virulence-associated genes (Zhang et al. 2004), single enzyme PFGE and EcoRI ribotyping. The strong epidemiological concordance of an MLVA protocol involving the assessment of eight TRs was also demonstrated by Volpe Sperry et al. (2008). Moreover, the experimental procedure was improved by the use of capillary gel electrophoresis for high throughput and accurate sizing. This protocol was also applied by Gorski et al. (2014) to study the effect of selective secondary enrichment on the biodiversity of the *L. monocytogenes* subtypes detected and evidence that a greater diversity of isolates was detected was provided. Another MLVA protocol coupled to high-resolution capillary electrophoresis was developed by Lindstedt et al. (2008) and applied to two unrelated sets of isolates.

This approach was based on the use of degenerate primers and was reported as slightly more discriminatory than PFGE for the first set Norwegian isolates (28 MLVA profiles and 24 PFGE profiles) and opposite for the second one (42 MLVA profiles and 43 PFGE profiles). Moreover, it was effectively applied for the characterization of isolates originating from three salmon-processing plants (Lunestad et al. 2013). The discriminatory power of the above mentioned protocols was evaluated and optimization strategies through the combination of tandem repeat sequences were proposed by Chen et al. (2011), Li et al. (2013) and Saleh-Lakha et al. (2013). In all cases improvement of the discrimination ability along with rapid and accurate subtyping was reported.

Clustered Regularly Interspaced Short Palindrome Repeats (CRISPRs)

The presence of short palindromic repeats of 25-50 nucleotides long that are interspaced by unique sequences of similar size has been recognized in both bacterial and archaeal genomes (van Belkum et al. 1998). These repeats are formed by embedding exogenously acquired nucleic acids into the chromosome through the CRISPR-associated proteins (Cas) that identify these foreign molecules and incorporate them into the CRISPR locus. The biological significance of this mechanism is yet to be fully elucidated; however a reasonable explanation may be the gain of some degree of immunity to the harmful foreign elements (Barrangou et al. 2007; Marraffini and Sontheimer, 2010; Bhaya et al. 2011; Horvath and Barrangou, 2010; van der Oost, 2009). Moreover, a role in virulence, infection and evasion of host immune system has also been suggested for an array of pathogenic bacteria including *L. monocytogenes* (Louwen et al. 2014; Sesto et al. 2014). The efficacy for microbial subtyping has been evaluated for several lactic acid bacteria (Horvath et al. 2009), *Yersinia pestis* (Pourcel et al. 2005), *Campylobacter jejuni* (Schouls et al. 2003), *Streptococcus pyogenes* (Hoe et al. 1999) and salmonellae (Bachmann et al. 2014; Pettengill et al. 2014). In the case of *L. monocytogenes* this approach has exhibited powerful ability for subtyping of the 1/2a (lineage II) and 1/2b (lineage I) serotypes. However, absence of typical CRISPR structures in many isolates of lineage I limited the subtyping efficacy (Di et al. 2014).

2.2. Biodiversity Studies

Several techniques have been applied for assessment of *L. monocytogenes* population biodiversity with arbitrary primed genotyping being the most frequently applied.

In Table 2 the primers and PCR conditions most frequently used for arbitrary primed clustering and differentiation of *L. monocytogenes* isolates are presented. The most critical aspect regarding this approach is the accurate standardization that is required for each individual step. Once this is achieved, this approach is a very powerful tool that may provide with optimal differentiation.

The discrimination power is typically quantified by the discrimination (D) index according to Hunter and Gaston (1988); each differentiation approach may provide with different discrimination index when applied to the same set of isolates and optimal differentiation may be achieved by their combination. The latter was recently demonstrated by Hadjilouka et al. (2014); with the proper combination of RAPD-PCR and rep-PCR it was possible to differentiate 17 *L. monocytogenes* isolates recovered from the same meat sample after the enrichment steps defined by ISO 11290-1:1996 (ISO, 1996).

Table 2. Primers and PCR conditions used for arbitrary primed clustering and differentiation of *L. monocytogenes* isolates

Primers (5'-3')	PCR conditions	Reference
PB1:GGAACTGCTA PB4: AAGGATCAGC HLWL74: ACGTATCTGC	$MgCl_2$, 2.5 mM; primer, 0.4 uM; dNTPs, 10 mM; *Taq* polymerase, 2 U 94°C/4 min, 35°C/2 min, 72°C/2 min, 43 cycles (94°C/1 min, 35°C/2 min, 72°C 2 min); 94°C/1 min, 35°C/2 min, 72°C/10 min	Byun et al. (2001)
OPM-01: GTTGGTGGCT	$MgCl_2$, 1.5 mM; primer, 1.38 uM; dNTPs, 0.2 mM; *Taq* polymerase, 1.25 U 44 cycles (94°C/1 min, 30°C/2, 72°C 2 min); 72°C/10 min	Lawrence et al. (1993)
M13: GAGGGTGGCGGTTCT	$MgCl_2$, 2.5 mM; primer, 4 uM; dNTPs, 0.2 mM; *Taq* polymerase, 1 U 95°C/2 min; 35 cycles (95°C/1 min, 38°C/1 min ramp to 72°C at 0.6°C/sec, 72°C 2 min); 72°C/10 min	Andrigetto et al. (2001)
UBC155: CTGGCGGCTG	$MgCl_2$, 2.5 mM; primer, 1.5 uM; dNTPs, 0.2 mM; *Taq* polymerase, 1 U. 95°C/2 min; 10 cycles (95°C/1 min, 45°C/1 min decreasing to 1 degree at each cycle to 36°C, 72°C 2 min), 30 cycles (95°C/1 min, 35°C/1 min, 72°C/2 min); 72°C/10min	Farber et al. (1994)
HLWL85: ACAACTGCTC	$MgCl_2$, 2.5 mM; primer, 4 uM; dNTPs, 0.2 mM; *Taq* polymerase, 1 U. 95°C/2 min; 45 cycles (95°C/1 min, 35°C/2 min, 72°C/1 min); 72°C/10min	Wernars et al. (1996)
$(GTG)_5$: GTGGTGGTGGTGGTG	$MgCl_2$,3mM; dNTP's, 0.2 mM; primer, 2uM; *Taq* polymerase, 2U 95°C/5 min; 30 cycles (95°C/30 sec, 40°C/1 min, 72°C/8 min); 72°C/16 min	Hadjilouka et al. (2014)

Another critical aspect regarding the application of these techniques is the cut-off similarity values, ranging from 75 to 95% that are very commonly used to assist interpretation of the results (Carminatti et al. 2004; Franciosa et al. 2001; Parisi et al. 2010; Zunabovic et al. 2012). Such values should be applied with great caution since they are designed to measure only genetic relatedness. If this is not taken into consideration, they may be misleading with regard to both biodiversity levels and conclusions concerning epidemiology. To avoid such misinterpretations it is better that two isolates are considered identical only when their genotypic profiles are identical.

3. Prevalence of Virulence-Associated Genes

Prevalence in meat and meat products has been reported to vary from 3.6%, 4.2%, 5.3% and 7.2% in processed, cured, smoked and roasted, and sauce pickled meat products, respectively, (Cordano and Rocourt, 2001; Wang et al. 2015) to 27%, 29%, 30% and 30.2% in bovine hides, ground beef, ready-to-eat poultry products and chicken samples, respectively (Khen et al. 2013; Fallah et al. 2012; Baek et al. 2000). *L. monocytogenes* has been isolated also from dairy products such as ice-cream, soft-cheese and cheese, the prevalence in which was reported at 3.48% (21/603), 0.78% (2/256) and 39.6% (46/116), respectively (Cordano and Rocourt, 2001; Montero et al. 2015). Prevalence studies in seafood have also revealed a significant presence of the pathogen in this food category. Occurrence of the pathogen has been reported to vary in fresh fish from 0.8% in marine fish samples to 43.3% in catfish fillet samples, in shellfish from 2% in mussel samples to 28.8% in fresh shrimp samples and in bivalves from 4% to 12.5%. In addition, smoked seafood samples have been positive for the presence of the pathogen with prevalence varying from 0.9% in smoked fish samples to 80.3% in cold-smoked salmon samples (Jami et al. 2014). Despite the fact that human food-borne listeriosis has been mainly linked to the consumption of poultry, beef, dairy products and seafood, the number of studies reporting incidence of *L. monocytogenes* in various fruits, vegetables, herbs and spices has been increased the last decade. Surveys on fresh products have reported a significant presence of *L. monocytogenes* and its prevalence has been reported to vary from 0.04% (2/50) in various juices to 36.8% (25/68) in radish (Sado et al. 1998; Francis et al. 1999). Generally it appears to be very variable and it can be as low as 0.3% (1/297), 0.7% (2/300) and 1.6% (1/63) in lettuce–iceberg samples, in fresh, minimally processed fruit, vegetables and sprout samples and in mixed salads, respectively, or even as high as 25.4% (88/347), in frozen vegetable salads and 33.7% in raw mixed salads (Lin et al. 1996; Abadias et al. 2008; U.S. FDA, 2013; Cordano and Jacquet, 2009; Kramarenko et al. 2013).

L. monocytogenes as a facultative intracellular bacterium is able to enter, survive and multiply in phagocytic and non-phagocytic cells, and to cross hosts' intestinal, blood-brain and fetoplacental barriers (Farber and Peterkin, 1991; Cossart and Lecuit, 1998). Shortly after ingestion, *L. monocytogenes* that ends up in host's stomach is exposed to proteolytic enzymes, a highly acidic environment, bile salts and nonspecific inflammatory attacks. Subsequently, the pathogen attaches to and enters host cells through passive and active actions activating its infection cycle that includes adhesion, invasion, escape from vacuole, intracellular multiplication and cell-to-cell spread (Roche et al. 2008).

The intracellular parasitism requires the expression of a series of genes, denoted as virulence genes due to their effect on the host. The last two decades, extensive investigations have identified many virulence-associated genes, the most notable of which are physically gathered in a 9-kb gene cluster, named Listeria Pathogenicity Island 1 (LIPI-1) consisting of the *prfA, plcA, hly, mpl, actA* and *plcB* genes.

Table 3. Primer sequences and respective amplicon sizes used for the *in vitro*, *in situ* and *in vivo* gene assays

Target gene	Primer sequences (5'-3')	Product (bp)	Reference
inlA	F:ACGAGTAACGGGACAAATGC R:CCCGACAGTGGTGCTAGATT	800	Liu et al. (2007)
	F:TGGTGTAAGATCGCT R:AAGTGATATAACTCC	478	Moreno et al. (2014)
	F:GGCTGGGCATAACCAAATTA R:CTTTTGTTGGTGCCGTAGGT	629	Montero et al. (2015)
inlB	F:TGGGAGAGTAACCCAACCAC R:GTTGACCTTCGATGGTTGCT	884	Liu et al. (2007)
	F:TGATGGAACGGTAAT R:TNATTTCTGTGCCCT	900	Moreno et al. (2014)
	F:CCTAAACCTCCGACCAAACA R:CCATTTCGCGCTTCTCTATC	293	Montero et al. (2015)
inlJ	F:TGTAACCCCGCTTACACAGTT R:AGCGGCTTGGCAGTCTAATA	238	Liu et al. (2007)
inlC	F:AATTCCCACAGGACACAACC R:CGGGAATGCAATTTTTCACTA	517	Liu et al. (2007)
	F:ATGCTAGTNTTAATTGTA R:CTAATTCTTGATAGGTTGTG	852	Moreno et al. (2014)
plcA	F:CTGCTTGAGCGTTCATGTCTCATCCCCC R:CATGGGTTTCACTCTCCTTCTAC	1484	Notermans et al. (1991)
	F:CTCGGACCATTGTAGTCATCTT R:CACTTTCAGGCGTATTAGAAACGA	326	Lomonaco et al. (2012)
	F:AACCATTATTATGCG R:TGCAGCATACTGACG	396	Moreno et al. (2014)
	F:TCCCATTAGGTGGAAAAGCA R:CGGGGAAGTCCATGATTAGA	840	Montero et al. (2015)
	F:CGAGCAAAACAGCAACGATA R:CCGCGGACATCTTTTAATGT	129	Jaradat et al. (2002)
plcB	F:CAGCTCCGCATGATATTGAC R:CTGCCAAAGTTTGCTGTGAA	723	Montero et al. (2015)
	F:GGGAAATTTGACACAGCGTT R:ATTTTCGGGTAGTCCGCTTT	261	Jaradat et al. (2002)
prfA	F:CTGTTGGAGCTCTTCTTGGTGAAGCAATCG R:AGCAACCTCGGTACCATATACTAACTC	1060	Notermans et al. (1991)
	F:ACCAATGGGATCCACAAGA R:CAGCTGAGCTATGTGCGAT	467	Bubert et al. (1999)
	F:CTGCTAACAGCTGAGCTATG R:GCTACCGCATACGTTATC	404	Moreno et al. (2014)
	F:ACCAATGGGATCCACAAGAA R:GCTTCCCGTTAATCGAAAAAT	330	Montero et al. (2015)
hlyA	F:GCAGTTGCAAGCGCTTGGAGTGAA R:GCAACGTATCCTCCAGAGTGATCG	456	Paziak-Domanska et al. (1999)
	F:TAACAACGCAGTAAA R:CGTAAGTCTCCGAGG	566	Moreno et al. (2014)

Target gene	Primer sequences (5'-3')	Product (bp)	Reference
	F:GTCTACCAATTGCGCAACAA R:TGGTGTTTCCCGGTTAAAAG	1100	Montero et al. (2015)
actA	F:CGGAGGTTCCGCAAAAGATG R:CCTCCAGAGTGATCGATGTT	234	Jaradat et al. (2002)
	F:CGCCGCGGAAATTAAAAAAAGA R:ACGAAGGAACCGGGCTGCTAG	839	Suarez and Vazquez-Boland, (2001)
	F:CCAAGCGAGGTAAATACGGGA R:GTCCGAAGCATTTACCTCTT	650	Lomonaco et al. (2012)
	F:AAACAGAAGAGCAGCCAAGC R:TTCACTTCGGGATTTTCGTC	571	Montero et al. (2015)
	F:GACGAAAATCCCGAAGTAA R:CTAGCGAAGGTGCTGTTTCC	268	Jaradat et al. (2002)
iap	F:ACAAGCTGCACCTGTTGCAG R:TGACAGCGTGTGTAGTAGCA	131	Furrer et al. (1991)
llsA	F:CGATTTCACAATGTGATAGGATG R GCACATGCACCTCATAAC	280	Clayton et al. (2014)

Furthermore, the last decade a number of studies that has been conducted in *L. monocytogenes* strains isolated from food and clinical samples, aim to determine a valuable assay for the characterization of the pathogenic potential of *L. monocytogenes* and the separation of pathogenic from nonpathogenic species, based on the presence or absence of these genes. In addition, other important proteins such as internalins, also characterized as virulence-associated, have been extensively studied.

Pathogen's active invasion in host cells is performed by a family of surface proteins known as internalins. InlA and InlB are recognized as the most important of the 25 internalins or internalin-like proteins in *L. monocytogenes*. InlA (encoded by *inlA*) enables pathogen's invasion into host's epithelial cells, due to the local cytoskeletal rearrangements caused by the reaction between InlA and adhension protein E-cadherin. InlB (encoded by *inlB*) recognizes its cellular receptor Met and its co-receptor Clq-R and thus facilitates *L. monocytogenes* internalize into hepatocytes, fibroblast and epithelioid cells. During the later stages of infection (i.e., post-intestinal) other internalins such as InlC and InlJ (or lmo2821) are considered to contribute to *L. monocytogenes* invasion (Roche et al. 2008). In addition, InlJ assists the successful passage through the intestinal barrier (Sabet et al. 2005).

In addition to the above internalins, InlC2, InlD, InlE, InlF and InlH have been suggested to participate in pathogen's virulence potential, without however a well understood virulence role (Orsi et al. 2011). InlC2, InlD, InlE, and InlF are considered helpful for an aspect of pathogenicity other than invasion (Dramsi et al. 1997), while InlG, InlH, InlE mutants have shown decrease in *in vivo* virulence, without however any effect on pathogen's admission into epithelial cells or on intracellular multiplication (Rafflesbauer et al. 1998).

inlA and *inlB* genes have been detected in all *L. monocytogenes* strains studied so far, isolated from human cases, food and environment (Liu et al. 2007; Orsi et al. 2011; Jamali et al. 2013; Moreno et al. 2014; Soni et al. 2014). It is noteworthy that *inlA* has been detected in all *L. monocytogenes* strains isolated from human cases, food and environment – in both pathogenic strains with capacity to cause mouse mortality and non-pathogenic strains- but not in *L. ivanovii*, *L. innocua*, *L. seeligeri*, *L. welshimeri and L. grayi* isolates, indicating that presence of *inlA* can confirm *L. monocytogenes* species (Liu et al. 2007). Despite the fact that these major virulence genes seem to be present in all *L. monocytogenes* isolates, some of them carry premature stop codons (PMSCs) that lead to virulence attenuation of the strains. Perhaps the best studied polymorphisms leading to virulence attenuation in *L. monocytogenes* strains are those in *inlA,* for which there have been observed at least 18 distinct mutations (Jacquet et al. 2004; Van Stelten et al. 2010; Orsi et al. 2011; Shen et al. 2013). Therefore, it should be noted that presence of these genes does not indicate internalization certainty for the pathogen. Primer sequences and respective amplicon sizes used for the *in vitro*, *in situ* and *in vivo* gene assays are presented in Table 3.

inlJ and *inlC* genes were present in all *L. monocytogenes* isolates studied by Soni et al. (2014), Jamali et al. (2013) and Moreno et al. (2014), and in all 52 *L. monocytogenes* strains analyzed by Lomonaco et al. (2012), except for one isolate that was negative for *inlC* presence. In a study conducted by Liu et al. (2007), both genes were detected in all 27 pathogenic strains with capacity to cause mortality in A/J mice, whereas the 6 naturally non-pathogenic strains were negative for their presence. Interestingly however, 2 strains positive only for the presence of *inlJ* and 1 strain positive only for *inlC* retained the ability to cause mouse mortality, suggesting that only simultaneous absence indicates avirulence. *inlJ* presence has been described as an indicator for the differentiation between virulent and avirulent strains. However, some strains of the lineage III seem to lack *inlJ* but retain the ability to cause mouse mortality via intraperitoneal inoculation (Liu et al. 2007). *inlC* has also been reported to be absent from some lineage III strains (Orsi et al. 2011). Shen et al. (2013) validated this statement indicating that among 97 *L. monocytogenes* isolates, 96 (99%) strains belonging to lineage I and II were positive for the presence of *inlJ* and *inlC*, while 1 strain belonging to lineage III was negative for the presence of both genes. Lineage IV strains seem also to lack *inlJ* (Orsi et al. 2011). It should be noted however, that despite the fact that presence of the *inlC* and/ or *inlJ* genes in a strain implies potential virulence and ability to cause mouse mortality via the intraperitoneal route, it does not necessarily indicate the certainty that strains harboring these genes will cause disease in humans via the conventional oral ingestion (Liu et al. 2007).

Regarding *inlC2, inlD* and *inlE* genes studies reported that they were present in isolates from lineage I and lineage II, whilst *inlG, inlH* and *inlF* genes were only detected among strains belonging to lineage II (Orsi et al. 2011). Furthermore, analysis of 3 lineage III isolates reported that none of them harbored *inlF* or *inlH* (Jia et al. 2007).

In order to determine the pathogenic potential of *L. monocytogenes* strains, the best characterized virulence factors along with internalins are the genes belonging to gene cluster LIPI-1 that contains the major *L. monocytogenes* virulence genes. Transcription of the genes present in the LIPI-1 is mainly controlled by the transcriptional activator PrfA, a member of the cyclic AMP (cAMP) receptor protein (CRP)/FNR family of bacterial regulators that is encoded by *prfA*. In addition, the expression of more than 140 genes is directly or indirectly influenced by PrfA, highlighting the importance of this molecule to *L. monocytogenes* homeostasis (Paramithiotis et al. 2014). *hlyA* encodes a cholesterol-binding, pore-forming toxin (listeriolysin O, LLO) essential for the effective escape from the phagocytic vacuoles, allowing the escape into the cytoplasm. However, recent studies are revisiting the role of LLO during infection revealing new insights into its action before bacterial entry (Hamon et al. 2012). It has been established although that absence of LLO essentially leads to avirulence (Kathariou et al. 1987; Cossart et al. 1989). LLO cooperates with two phospholipases C, encoded in LIPI-1: PlcA and PlcB. The former is a phosphatidylinositol phospholipase C (PI-

PLC) encoded by *plcA*. Expression of *plcA* has been linked to pathogenic potential of *L. monocytogenes* and *L. ivanovii* and is considered a reliable marker for discrimination between pathogenic and non-pathogenic *Listeria* spp. (Aurora et al. 2008). PlcB is a phosphatidylcholine phospholipase C (PC-PLC), a broad spectrum phospholipase C encoded by *plcB*. PC-PLC in collaboration with PI-PLC and LLO promote lysis of the primary vacuole. In addition, PC-PLC assists LLO promote lysis of the double-layer membrane vacuoles (Roche et al. 2008). However, PlcB can also provoke lysis of the primary vacuole in the absence of LLO (Grundling et al. 2003). *actA* encodes the protein ActA, a surface protein that appears to be a multifunctional virulence factor. It is necessary for actin-based intra-and inter-cellular motility, playing a major role in cell-to-cell spread and in epithelial cell invasion.

The last decade, studies conducted for the determination of the pathogenic potential of *L. monocytogenes* strains, in order to draw conclusions, target all the above genes along with *iap* gene, in each isolate. *iap* gene, also termed *p60*, encodes the invasion associated protein Iap which is suggested to play a major role in maintaining the invasive phenotype in mouse fibroblasts, hepatocytes and macrophages. In addition, it has been suggested as necessary for the proper cell division (Bubert et al. 1992; Hess et al. 1995).

Presence of these genes has been evaluated in *L. monocytogenes* strains isolated from various samples, such as vegetables and soil samples (Soni et al. 2014), cheese samples (Lomonaco et al. 2012), pork slaughter houses, markets and human infection samples (Moreno et al. 2014), patients with spontaneous abortions (Kaur et al. 2007) and apparently healthy animals (Yadav et al. 2011). Primer sequences and respective amplicon sizes used for the gene assays are presented in Table 3. Results of the above studies reported the presence of *plcA, prfA, acta, hlyA* and *iap* genes in vegetable and soil samples, presence of *plcA, plcB, prfA* and *hlyA* genes in pork slaughter houses, markets and human infection samples and presence of *plcA, plcB, hlyA, prfA* and *actA* but absence of *iap* gene in a *L. monocytogenes* strain isolated from an apparently healthy animal. However, a study conducted in 4 strains, isolated from bovine mastitis cases, indicated presence of the five virulence-associated genes *plcA, prfA, actA, hlyA* and *iap* only in the strain that was found to be pathogenic. The remaining three isolates were hemolytic but negative for *plcA* gene, therefore were characterized as non- pathogenic. All of these strains harbored *hlyA* and in addition each of them harbored *prfA*, *actA* and *iap*, and *iap*, respectively (Rawool et al. 2007). Similarly to these results, Kaur et al. (2007) reported that in spite of the haemolytic activity expressed by all 4 *L. monocytogenes* strains, isolated by patients with spontaneous abortions, only the two strains characterized as pathogenic possessed all the five virulence-associated genes *prfA, plcA, hlyA, actA* and *iap*. The remaining two non-pathogenic were negative for the presence of all the genes and of *plcA* and *prfA*, respectively. Presence of the above virulence-associated genes suggests pathogenic potential of the isolates and a risk to human health, although strains positive for these genes are not certainly pathogenic, as these genes are normally present in *L. monocytogenes* (virulent, less virulent and avirulent), However, strains that do not harbor these genes are likely to be avirulent (Lomonaco et al. 2012).

The characteristic hemolytic phenotype of *L. monocytogenes* has been attributed to the activity of LLO, encoded by the *hly* gene that is located within the LIPI-1. Nevertheless, it has been disclosed that several *L. monocytogenes* strains of lineage I possess an additional pathogenicity island termed LIPI-3 that encodes a second hemolysin, called listeriolysin S (LLS) (Clayton et al. 2011). LLS is a post-translationally modified cytolysin and is a member of the broader family of thiazole/oxazole modified microcins (TOMMs). This hemolysin is not normally expresses *in vitro* and *hly* mutants give a non-hemolytic phenotype on blood agar (Clayton et al. 2014). It has been established although, that this hemolysin contributes to the survival of the pathogen in polymorphonuclear leukocytes as well as to virulence in the murine model (Cotter et al. 2008).

L. monocytogenes strains originated from human, animal, food, feed and sewage were screened for the presence of LIPI-3 and *lls* genes. Obaidat et al. (2015) reported a prevalence of 28% and 40% in 1/2b and 4b strains isolated from fish samples, respectively, while Althaus et al. (2014) reported a prevalence of 87.8% in lineage I strains that had been isolated from patients. Furthermore, 27 of the 46 (59%) *L. monocytogenes* isolates belonging to lineage I were positive of LIPI-3 presence, contrary to the 23 lineage II and 5 lineage III isolates in which LIPI-3 was absent (Cotter et al. 2008). Among 83 *L. monocytogenes* isolates (30 lineage I, 50 lineage II and 3 lineage III) the *lls* gene was detected in 21 of the 30 (70%) *L. monocytogenes* lineage I strains and in all cases the full complement of LIPI-3 genes was present. However, *llsA* gene was not detected in any lineage II or lineage III stain (Clayton et al. 2014). Contrary to these results, Shen et al. (2013) reported twelve *llsX*-positive isolates, eleven belonging to lineage I and one belonging to lineage III.

4. Transcriptomic Responses on Food Matrixes

Response to an environmental stress necessary includes three steps:

- stress sensing-signaling
- activation of response mechanisms

Changes in the environment are sensed by a two-component regulatory system consisting of *lisR* and *lisK* (Cotter et al. 1999). Moreover, it has been reported that *htrA*, a gene described as necessary for osmotolerance (Wonderling et al. 2004) as well as virulence potential (Stack et al. 2005) is under the transcriptional control of this system (Sleator and Hill, 2005).

Stress response involves the activation of several mechanisms. Some of them may be considered to belong to a core mechanism, such as uptake and accumulation of compatible solutes, the DEAD-box proteins and the SOS response, the activation of which is useful in combating multiple stresses, and some of them are stress-dependent. In the following paragraphs an update is offered regarding the research that has taken place regarding their occurrence and regulation in *L. monocytogenes*.

Sigma B

Sigma B is a transcriptional regulator that possesses a central role in virulence as well as stress response. Regarding the latter, it is conserved among several Gram positive bacteria contributing to their adaptation to stressful conditions (van Schaik and Abee, 2005). The importance for survival on acid stress, high hydrostatic pressure and freeze (Wemekamp-Kamphuis et al. 2004a) in the resistance of planktonic and biofilm-forming cells in both static and continuous-flow systems to benzalkonium chloride and peracetic acid (van der Veen and

Abee, 2010) has already been described. Interestingly, the contribution to survival seems to take place in a strain-dependent manner (Oliver et al. 2010). Moreover, sigma C, sigma H, sigma L as well as other transcriptional regulators such as CtsR and HrcA also play important and similar roles. All these regulons along the one controlled by PrfA form a regulatory network through specific overlaps that provide the cell with regulatory redundancies and the ability to accurately control gene expression to fit the environmental changes (Chaturongakul et al. 2011; Mujahid et al. 2013; Guariglia-Oropeza et al. 2014).

Uptake and Accumulation of Compatible Solutes

Uptake and intracellular accumulation of compatible solutes is one of the adaptation mechanisms regarding cold and osmotic stress. In the case of the former, glycine betaine and carnitine are the major cryoprotective osmolytes. Although *L. monocytogenes* does not seem to possess the mechanism for their biosynthesis, they can be abundantly recovered at high levels in foods of both plant and animal origin. Presence of sigma B promoter sites upstream of all respective operons supports the sigma B dependence of their upregulation during cold stress (Wemekamp-Kamphuis et al. 2004b). However, presence of another promoter site upstream to *gbuABC* as well as a putative sigma A binding site upstream to *betL* have also been suggested (Sleator et al. 1999; Cetin et al. 2004). Short peptides may also play the role of compatible solute. Their accumulation has been suggested to occur through the oligopeptide permease (Opp) transport system (Borezee et al. 2000).

Regarding the osmotic stress, glycine betaine, proline betaine, acetyl carnitine, carnitine, y-butyrobetaine and 3-dimethylsulphoniopropionate function as osmoprotectants (Bayles and Wilkinson, 2000). In the presence of these compounds the growth rate of salt-stressed cells was reported as significantly improved (Bayles and Wilkinson, 2000).

DEAD-Box Proteins – SOS Response

DEAD-box proteins are conserved RNA helicases that participate in many cellular processes (Linder and Jankowsky, 2011). Their significance in tolerance of various stresses has been the subject of some study. A single insertion and concomitant inactivation of LMOf2365_1746 that encodes a cold-induced putative RNA helicase was held responsible for failure of a *L. monocytogenes* strain to grow at low temperatures (Azizoglu and Kathariou, 2010). More recently the putative DEAD-box RNA helicase genes lmo0866, lmo1450 and lmo1722 were reported as necessary for cold tolerance and motility and the two former to possess important roles in tolerance of *L. monocytogenes* EGD-e to ethanol, heat, alkali, and oxidative stresses (Markkula et al. 2012a; b).

The SOS response is induced upon exposure to stress that may cause damage to DNA. This response directs DNA repair and increases mutation rates in an attempt to improve chances of survival (van der Veen et al. 2010a).

The SOS response in *L. monocytogenes* was characterized by van der Veen et al. (2010a); it was reported to consist of 29 genes in 16 operons regulated by LexA encoding translesion DNA polymerases, DNA-repair proteins as well as a bile resistance system. Moreover, RecA was also reported to possess important role in stress survival and mutagenesis (Zgur-Bertok, 2013).

Heat Shock Response

Exposure to higher temperatures induces the expression of a set of proteins that are collectively referred to as heat shock proteins (HSPs). These include molecular chaperones (e.g., DnaK, GroES, GroEL) that attempt refolding of the proteins being damaged by stress conditions and ATP-dependent proteases (e.g., Clp proteins) that degrade the proteins whose refolding is not possible. Regarding Gram-positive bacteria, at least three classes of genes have been described as heat inducible (Yura and Nakahigashi, 1999):

- Class I encodes the chaperones DnaKJ-GrpE and GroEL-GroES the expression of which is negatively affected by *hrcA*,
- Class II is an extended regulon controlled by sigma B and is activated by a variety of environmental stimuli and
- Class III that includes the remaining heat inducible genes such as the *clpC* operon.

An integrated heat shock response of *L. monocytogenes* was studied by van der Veen et al. (2007) using DNA microarrays. Apart from genes previously described as part of the heat-shock regulon, numerous other genes were differentially expressed. More accurately, genes belonging to the SOS response and DNA repair were upregulated whereas downregulation was reported for cell wall hydrolases and genes involved in cell division. Mixed response was reported for other cell wall as well as virulence associated genes.

Regarding class I HSPs, expression of the *groESL* operon is induced during intracellular infection, while DnaK is required for efficient phagocytosis with macrophages (Gahan et al. 2001; Hanawa et al. 1999). The latter has been reported to participate in a number of cellular processes (Gething and Sambrook, 1992; Hartl, 1996). A role in pathogenicity was also indicated since it was reported to be involved in the transcription of *flaA* and *lmaB* (Hanawa et al. 2002); the former encodes for a flagellin with murein-hydrolyzing activity and the latter for antigen B.

As far as class III HSPs were concerned, expression of *clpC*, *clpE* and *clpP* seems to be controlled by CtsR and this regulon is very similar to the one of *B. subtilis* (Nair et al. 2000b). The role of Clp proteins in intracellular parasitism has been studied to some extent. Listeriolysin O activity was reported as significantly reduced in a *clpP* mutant under stress conditions and fully restored by complementation of the mutant Gaillot et al. (2000). ClpC has been reported to promote early escape from the phagosomal compartment of macrophages (Rouquette et al. 1996; 1998).

Furthermore, evidence that *clpC* expression is negatively controlled by PrfA were provided by Ripio et al. (1998). *clpE* expression on the contrary is not regulated by environmental stimuli such as elevated temperatures or salt stress but is upregulated in the absence of ClpC while synthesis of the latter is not affected by *clpE* deletion (Nair et al. 1999).

Moreover ClpC is involved in the invasion of hepatocytes *in vivo* during infection most probably due to the ClpC-dependent modulation of the invasion virulence factors InlA, InlB, and ActA (Nair et al. 2000a). *clpB*, another member of the CtsR regulon was described by Chastanet et al. (2004). A role in induced thermotolerance as well as pathogenicity but no involvement in salt, detergent, puromycin or cold stress tolerance was indicated.

Cold Shock Response

Exposure to a cold shock results in the induction of 12 proteins referred to as Cold Shock Proteins (CSPs) whereas balanced growth at low temperatures results in the induction of only 4 that are referred to as Cold Acclimation Proteins (CAPs) and are part of CSPs (Bayles et al. 1996). However, another study revealed the induction of 38 CSPs in *L. monocytogenes* EGD-e strain (Phan-Thanh and Gormon, 1995). Currently, research is focus on proteins belonging to the CspA family, i.e., proteins characterized by the presence of a conserved domain highly homologous to the cold shock domain of eukaryotic Y-box proteins (Graumann and Marahiel, 1998; Sommerville and Ladomery, 1996). In the case of *L. monocytogenes* strains EGD-e and F2365 seem to harbor three proteins belonging to that family, namely CspA, CspB and CspD (Nelson et al. 2004). The importance of these proteins to both cold and osmotic (NaCl) stress adaptation was highlighted by Schmid et al. (2009). Depending on the type of stress their functional importance and concomitantly contribution to adaptation and survival seemed to be different; CspA>CspD>CspB for cold stress and CspD>CspA/CspB for osmotic stress inflicted by NaCl.

Durack et al. (2013) studied the transcriptomic response to hyperosmotic and low temperature conditions and reported the existence of global stress-adaptation mechanisms. More accurately, genes associated with translation, transcription, peptidoglycan and fatty acid biosynthesis, transport across the cell membrane as well as cell division were upregulated. On the other hand, genes involved in carbohydrate metabolism, flagella assembly as well as belonging to sigma B and PrfA regulons were downregulated.

Acid Shock Response

Acid stress survival is obtained by the glutamate decarboxylase (GAD) system the arginine deiminase (ADI) system and the F_0F_1ATPase enzyme complex. The former consists of three genes, namely *gadA*, *gadB* and *gadC* genes. The two former encode two glutamate decarboxylases and the latter a glutamate/y-aminobutyrate antiporter (Cotter et al. 2001). Glutamate is taken up by the cell through a specific transporter, decarboxylated intracellularly to y-aminobutyrate and exported via a cell membrane located antiporter. Through this procedure a proton is utilized resulting in the increase of the pH value. The arginine deiminase system consists of arginine deiminase encoded by *arcA*, catabolic ornithine carbamoyltransferase encoded by *arcB*, arginine-ornithine antiporter encoded by *arcD* and carbamate kinase encoded by *arcC*. *argR* has been characterized as the regulator of this system. Arginine is transported intracellularly, deiminated to ornithine that is then transported extracellularly. For each molecule of arginine converted to ornithine two moles of ammonia and one of ATP are produced. The former may react with protons resulting in pH value increase whereas the latter may be used for biosynthetic processes or to remove protons via the F_0F_1ATPase enzyme complex (Stack et al. 2008). The operon of the latter consists of nine genes and was characterized by Cotter et al. (2000).

The contribution of *fri*, a gene encoding a ferritin-like protein, in acid stress tolerance was recently reported by Milecka et al. (2015).

Osmotic Response

The osmotic response is very similar to the cold shock response. Salt Shock Proteins (SSPs) and Salt Acclimation Proteins (SAPs) were induced upon exposure to the stress and balanced growth, respectively. DnaK and Ctc were identified among the former and GbuA among the latter (Duche et al. 2002a). Moreover, proteins such as RelA, HtrA, KdpE, LisRK, ProBA and BtlA have also been reported to facilitate growth of *L. monocytogenes* under osmotic stress (Sleator and Hill, 2005; Wonderling et al. 2004; Okada et al. 2002; Brondsted et al. 2003; Kallipolitis and Ingmer, 2001; Sleator et al. 2001a; b; Begley et al. 2003).Growth and global gene expression of *L. monocytogenes* strain F2365 under salt stress was studied by Bae et al. (2012) and reported upregulation of genes involved in the uptake of glycine betaine/L-proline and downregulation of genes associated with a putative phosphoenolpyruvate-dependent sugar phosphotransferase system (PTS), metabolic enzymes, and virulence factors.

Response to Antimicrobials

Benzalkonium chloride (BAC) is a disinfectant based on quaternary ammonium compounds (QAC) that is widely used in the food industry for surface sanitation. Its mode of action includes damage on membrane permeability that results in leakage of cellular material (McDonnell and Russell, 1999). However, several factors such as food debris may result in the reduction of the effectiveness leading to an acquired resistance. The latter has been the subject of extended study. Presence of a putative resistance cassette, brcABC, either on the chromosome or on a large plasmid (pLM80) in several strains as well as its induction upon sublethal BAC concentrations have been reported (Elhanafi et al. 2010; Dutta et al. 2013). Furthermore, exposure to BAC induces a multidrug resistance efflux pump (MdrL) (Romanova et al. 2006) as well as upregulation of sigma B (van der Veen et al. 2010b). In a recent study by Tamburro et al. (2015) involving food and clinical isolates, a significant association between BAC exposure and *mdr*L overexpression as well as no correlation with *lde* expression were reported. Moreover, the importance of strain variability was highlighted.

Carnocycln A is a 5.9-kDa cyclic, class IIc bacteriocin produced by *Carnobacterium maltaromaticum* that exhibits strong antilisterial activity (Martin-Visscher et al. 2008). Liu et al. (2014) studied the response of *L. monocytogenes* 08-5923 to the exposure to this peptide for 15 min, 4 and 30 h. Upregulation of genes involved in metabolism (*baiE*, *trn*, and *pykA*), cell wall synthesis (*murZ* and *dacB2*), and cell division (*clpE* and *divIVA*) were reported after 15 min while downregulation of genes involved in cell division, cell wall synthesis, flagellar synthesis, and metabolism was observed after 4 h. Finally, absence of flagella was observed following 30 h of exposure.

The transcriptomic response of several *L. monocytogenes* strains after exposure to the growth supernatant of pediocin producing *Lb. plantarum strain* WHE 92 for 10, 60 and 180 min was studied by Laursen et al. (2015). Temporal expression changes in genes regulated by LisRK and the alternative sigma factors SigB and SigL was observed. Additionally, upregulation of several genes involved in bacteriocin resistance was reported.

Chlorine dioxide is a sanitizer that is increasingly used in the food industry to reduce the microbial load of food-contact surfaces and fresh produce (Lopez-Galvez et al. 2010; Han et al. 2001; Ryu and Beuchat, 2005; Mahmoud et al. 2008). The transcriptional response of early log phase a *L. monocytogenes* strain was studied by Pleitner et al. (2014). A total of 340 of genes were reported to be differentially expressed, of which 113 and 16 belonged to sigma B and CtsR regulons, respectively, highlighting their importance in surviving sublethal concentrations of the sanitizer.

Response to Actual Food Matrix

The response during growth on an actual food matrix has been studied only to a limited extend and mostly regarding virulence genes. Liu and Ream (2008) monitored gene expression of *L. monocytogenes* strain F2365 after 24 h of growth at 4°C in UHT skim milk. Upregulation was reported for 26 genes encoding for transport and binding proteins, transcriptional regulators, proteins in amino acid biosynthesis and energy metabolism, protein synthesis and cell division while downregulation was reported for 14 genes encoding for transport and binding proteins, protein synthesis, cellular processes, cell envelope, energy metabolism, and one transcriptional regulator.

The expression of *prfA*, *inlA*, *sigB* and *clpC* of three *L. monocytogenes* strains after incubation for 48h at 7°C of standard liver pâté as well as three variants with reduced NaCl content, one of which has also been supplied with organic acids (Ca-acetate and Ca-lactate) was assessed by Olesen et al. (2010). Strain-dependent effects were reported; more accurately a reduction in *prfA* for one strain and a reduction in *inlA* and *sigB* expression for two strains were reported when grown in standard liver pâté compared to growth in BHI broth. Reduction of NaCl content of the standard liver pâté did not change relative transcription levels of *prfA*, *inlA*, *sigB* or *clpC* (except for *prfA* in one strain and *sigB* in another).

Eleven *L. monocytogenes* strains were selected by Alessandria et al. (2013) in a comparative study on the expression of *plcA*, *hly*, *iap* and *sigB* in synthetic laboratory medium and in a soft cheese at two temperatures (4 and 12°C) and for different times (24 and 48 h). Significant expression differences emerged at 12°C after 48 h. The results obtained highlighted once again the differential expression in different strains, underlining the different pathogenic potentials among different isolates. The differences observed could not be associated neither with the origin of the strains nor their serotype.

Bae et al. (2011) studied the transcriptomic response of *L. monocytogenes* strain F2365 after growth on RTE deli slices of smoked turkey breast for 5 d at 15°C. A set of 84 genes involved in energy metabolism, fatty acid and phospholipid metabolism, biosynthesis of proteins, transport and binding proteins, DNA metabolism, cellular processes, and regulatory functions were transcriptionaly regulated. Regarding *sigB*, *prfA*, *inlA*, *inlB*, *plcA*, *plcB*, and *hly*, no significant differences were observed, compared to growth in BHI broth.

Recently, Mataragas et al. (2015) reported on differential gene expression profiling of *L. monocytogenes* in Cacciatore and Felino salami. A cocktail of five *L. monocytogenes* strains was used to inoculate the batter of each fermented sausage that was then fermented according to established practices. *lmo0669*, *gbuA* and *lmo1421* were up-regulated during the ripening of Cacciatore; the two latter were also upregulated during the ripening of Fellino as well. *sigB* expression was induced in both sausages throughout the whole process while *prfA* was down-regulated during the fermentation of Cacciatore.

References

Aarts, HJM; Hakemulder, LE; Van Hoef, AMA. Genomic typing of *Listeria monocytogenes* strains by automated laser fluorescence analysis of amplified fragment length polymorphism fingerprint patterns. *Int. J. Food Microbiol.,* 1999, 49, 95–102.

Abadias, M; Usall, J; Anguera, M; Solsona, C; Vinas I. Microbiological quality of fresh, minimally-processed fruits and vegetables, and sprouts from retail establishments. *Int. J. Food Microbiol.*, 2008, 123(1-2), 121-9.

Alessandria, V; Rantsiou, K; Dolci, P; Zeppa G; Cocolin, L. A comparison of gene expression of *Listeria monocytogenes in vitro* and in the soft cheese Crescenza. *Int. J. Dairy Technol.*, 2013, 66, 83-89.

Allix, C; Supply, P; Fauville-Dufaux, M. Utility of fast mycobacterial interspersed repetitive unit-variable number tandem repeat genotyping in clinical mycobacteriological analysis. *Clin. Infect. Dis.*, 2004, 39, 783-789.

Althaus, D; Lehner, A; Brisse, S; Maury, M; Tasara, T; Stephan, R. Characterization of *Listeria monocytogenes* strains isolated during 2011–2013 from human infections in Switzerland. *Foodborne Pathog. Dis.*, 2014, 11, 753–758.

Andrigetto, C; Zampese, L; Lombardi, A. RAPD-PCR characterization of lactobacilli isolated from artisanal meat plants and traditional fermented sausages of Veneto region (Italy). *Lett. Appl. Microbiol.*, 2001, 33, 26–30.

Aurora, R; Prakash, A; Prakash, S; Rawool, DB; Barbuddhe, SB. Comparison of PI-PLC based assays and PCR along with in vivo pathogenicity tests for rapid detection of pathogenic *Listeria monocytogenes*. *Food Control*, 2008, 19, 641–647.

Azizoglu, RO; Kathariou, S. Inactivation of a cold-induced putative RNA helicase gene of *Listeria monocytogenes* is accompanied by failure to grow at low temperatures but does not affect freeze-thaw tolerance. *J. Food Prot.*, 2010, 73 (8), 1474–1479.

Bachmann, NL; Petty, NK; Ben Zakour, NL; Szubert, JM; Savill J; Beatson, SA. Genome analysis and CRISPR typing of *Salmonella* enterica serovar Virchow. *BMC Genomics*, 2014, 15, 389.

Bae, D; Crowley, MR; Wang, C. Transcriptome analysis of *Listeria monocytogenes* grown on a ready-to-eat meat matrix. *J. Food Prot.*, 2011, 74, 1104–1111.

Bae, D; Liu, C; Zhang, T; Jones, M; Peterson, SN; Wang, C. Global gene expression of *Listeria monocytogenes* to salt stress. *J. Food Prot.*, 2012, 75 (5), 906–912.

Baek, SY; Lim, SY; Lee, DH; Min, KH; Kim, CM. Incidence and characterization of *Listeria monocytogenes* from domestic and imported foods in Korea. *J. Food Prot.*, 2000, 63, 186-189.

Barrangou, R; Fremaux, C; Deveau, H; Richards, M; Boyaval, P; Moineau, S; Romero, DA; Horvath, P. CRISPR provides acquired resistance against viruses in prokaryotes. *Science*, 2007, 315, 1709-1712.

Barrett, TJ; Gerne-Smidt, P; Swaminathan, B. Interpretation of pulsed-field gel electrophoresis patterns in foodborne disease investigations and surveillance. *Foodborne Pathog. Dis.*, 2006, 3, 20-31.

Bayles, DO; Annous, BA; Wilkinson, BJ. Cold stress proteins induced in *Listeria monocytogenes* in response to temperature downshock and growth at low temperatures. *Appl. Environ. Microbiol.*, 1996, 62 (3), 1116–1119.

Bayles, DO; Wilkinson, BJ. Osmoprotectants and cryoprotectants for *Listeria monocytogenes*. *Lett. Appl. Microbiol.*, 2000, 30 (1), 23–27.

Begley, M; Hill, C; Gahan, CGM. Identification and disruption of *btlA*, a locus involved in bile tolerance and general stress resistance in *Listeria monocytogenes*, *FEMS Microbiol. Lett.*, 2003, 218, 31.

Bergamini, F; Iori, A; Massi, P; Pongolini, S. Multilocus variable-number of tandem-repeats analysis of *Salmonella enterica* serotype Gallinarum and comparison with pulsed-field gel electrophoresis genotyping. *Vet. Microbiol.*, 2011, 149, 430-436.

Bhaya, D; Davison, M; Barrangou, R. CRISPR-Cas systems in bacteria and archaea: versatile small RNAs for adaptive defense and regulation. *Ann. Rev. Genet.*, 2011, 45, 273-297.

Borezee, E; Pellegrini, E; Berche, P. OppA of *Listeria monocytogenes,* an oligopeptide-binding protein required for bacterial growth at low temperature and involved in intracellular survival. *Infect. Immun.*, 2000, 68, 7069–7077.

Boxrud, D; Pederson-Gulrud, K; Wotton, J; Medus, C; Lyszkowicz, E; Besser, J; Bartkus, JM. Comparison of Multiple-Locus Variable-Number Tandem Repeat Analysis, Pulsed-Field Gel Electrophoresis, and Phage Typing for subtype analysis of *Salmonella enterica* serotype Enteritidis. *J. Clin. Microbiol.*, 2007, 45, 536-543.

Brody, MS; Price, CW. *Bacillus licheniformis sigB* operon encoding the general stress transcription factor sigma B. *Gene,* 1998, 212, 111–118.

Brondsted, L; Kallipolitis, BH; Ingmer, H; Knochel, S. *kdpE* and a putative RsbQ homologue contribute to growth of *Listeria monocytogenes* at high osmolarity and low temperature, *FEMS Microbiol. Lett.*, 219, 233, 2003.

Bubert, A; Kuhn, M; Goebel, W; Köhler, S. Structural and functional properties of the p60 proteins from different *Listeria* species, *J. Bacteriol.*, 1992,174, 8166-8171.

Bubert, A; Sokolovic, Z; Chun, SK; Papatheodorou, L; Simm, A; Goebel, W. Differential expression of *Listeria monocytogenes* virulence genes in mammalian host cells. *MGG.,* 1999, 261, 323–336.

Byun, SK; Jung, SC; Yoo, HS. Random amplification of polymorphic DNA typing of *Listeria monocytogenes* isolated from meat. *Int. J. Food Microbiol.*, 2001, 69, 227–235.

Cantinelli, T; Chenal-Francisque, V; Diancourt, L; Frezal, L; Leclercq, A; Wirth, T; Lecuit, M; Brisse, S. “Epidemic Clones” of *Listeria monocytogenes* are widespread and ancient clonal groups. *J. Clin. Microbiol.*, 2013, 51 (11), 3770–3779.

Carminati, DA; Perrone, G; Giraffa, E; Neviani, Mucchetti, G. Characterization of *Listeria monocytogenes* strains isolated from Gorgonzola cheese rinds. *Food Microbiol.,* 2004, 21, 801–807.

Cetin, MS; Zhang, C; Hutkins, RW; Benson, AK. Regulation of transcription of compatible solute transporters by the general stress sigma factor, sigma B, in *Listeria monocytogenes. J. Bacteriol.*, 2004, 186, 794–802.

Chastanet, A; Derre, I; Nair, S; Msadek, T. *clpB*, a novel member of the *Listeria monocytogenes* CtsR regulon, is involved in virulence but not in general stress tolerance. *J. Bacteriol.*, 2004, 186 (4), 1165–1174.

Chaturongakul, S; Raengpradub, S; Palmer, ME; Bergholz, TM; Orsi, RH; Hu, Y; Ollinger, J; Wiedmann, M; Boor, KJ.. Transcriptomic and phenotypic analyses identify coregulated, overlapping regulons among PrfA, CtsR, HrcA, and the alternative sigma factors B, C, H, and L in *Listeria monocytogenes. Appl. Environ. Microbiol.*, 2011, 77, 187–200.

Chen, S; Li, J; Saleh-Lakha, S; Allen, V; Odumeru, J. Multiple-locus variable number of tandem repeat analysis (MLVA) of *Listeria monocytogenes* directly in food samples. *Int. J. Food Microbiol.*, 2011, 148, 8–14.

Chen, Y; Zhang, W; Knabel, SJ. Multi-Virulence-Locus Sequence typing clarifies epidemiology of recent listeriosis outbreaks in the United States. *J. Clin. Microbiol.,* 2005, 43 (10), 5291–5294.

Chen, Y; Zhang, W; Knabel, SJ. Multi-virulence-locus sequence typing identifies single nucleotide polymorphisms which differentiate epidemic clones and outbreak strains of *Listeria monocytogenes. J. Clin. Microbiol.*, 2007, 45, 835-846.

Clayton, EM; Daly, KM; Guinane, CM; Hill, C; Cotter PD; Ross, PR. Atypical *Listeria innocua* strains possess an intact LIPI-3. *BMC Microbiol.,* 2014, 14, 58. http://www.biomedcentral.com/1471-2180/14/58.

Clayton, EM; Hill, C; Cotter, PD; Ross, RP. Real-time PCR assay to differentiate listeriolysin S-positive and -negative strains of *Listeria monocytogenes. Appl. Environ. Microbiol.*, 2011, 77, 163–171.

Cordano, AM; Rocourt, J. Occurrence of *Listeria monocytogenes* in food in Chile. *Int. J. Food Microbiol.,* 2001, 70, 175-178.

Cordano, AM; Jacquet, C. *Listeria monocytogenes* isolated from vegetable salads sold at supermarkets in Santiago, Chile: Prevalence and strain characterization. *Int. J. Food Microbiol.*, 2009, 132, 176–179.

Cossart, P; Lecuit, M. Interactions of *Listeria monocytogenes* with mammalian cells during entry and actin-based movement: Bacterial factors, cellular ligands and signaling, *EMBO J.,* 1998, 17, 3797.

Cossart, P; Vicente, MF; Mengaud, J; Baquero, F; Perez-Diaz, JC; Berche, P. Listeriolysin O is essential for virulence of *Listeria monocytogenes*: Direct evidence obtained by gene complementation. *Infect. Immun.*, 1989, 57, 3629–3636.

Cotter, PD; Draper, LA; Lawton, EM; Daly, KM; Groeger, DS; Casey, PG; Ross, RP; Hill, C. Listeriolysin S, a novel peptide haemolysin associated with a subset of lineage I *Listeria monocytogenes. PLOS Pathog.*, 2008, 4:e1000144.

Cotter, PD; Emerson, N; Gahan, CG; Hill, C. Identification and disruption of lisRK, a genetic locus encoding a two-component signal transduction system involved in stress tolerance and virulence in *Listeria monocytogenes. J. Bacteriol.*, 1999, 181 (21), 6840–6843.

Cotter, PD; Gahan, CG; Hill, C. A glutamate decarboxylase system protects *Listeria monocytogenes* in gastric fluid. *Mol. Microbiol.,* 2001, 40 (2), 465–475.

Cotter, PD; Gahan, CGM; Hill, C. Analysis of the role of the *Listeria monocytogenes* F0F1ATPase operon in the acid tolerance response, *Int. J. Food Microbiol.*, 2000, 60, 137.

Dass, SC; Abu-Ghannam, N; Antony-Babu, S; Cummins, EJ. Ecology and molecular typing of *L. monocytogenes* in a processing plant for cold-smoked salmon in the Republic of Ireland. *Food Res. Int.*, 2010, 43, 1529–1536.

Davis, MA; Baker, KNK; Call, DR; Warnick, LD; Soyer, Y; Wiedmann, M; Grohn, Y; McDonough, PL; Hancock, DD; Besser, TE. Multilocus Variable-Number Tandem-Repeat method for typing *Salmonella enterica* serovar Newport. *J. Clin. Microbiol.*, 2009, 47, 1934-1938.

Di, H; Ye, L; Yan, H; Meng, H; Yamasak, S; Shi, L. Comparative analysis of CRISPR loci in different *Listeria monocytogenes* lineages. *Biochem. Biophys. Res. Commun.*, 2014, 454, 399–403.

Doulgeraki, AI; Paramithiotis, S; Kagkli, DM; Nychas, GJE. Lactic acid bacteria population dynamics during minced beef storage under aerobic or modified atmosphere packaging conditions. *Food Microbiol.,* 2010, 27, 1028-1034.

Doulgeraki, AI; Paramithiotis, S; Nychas, GJE. Development of *Enterobacteriaceae* community during storage of minced beef under aerobic or modified atmosphere packaging conditions. *Int. J. Food Microbiol.,* 2011, 145, 77-83.

Dramsi, S; Dehoux, P; Lebrun, M; Goossens, PL; Cossart, P. Identification of four new members of the internalin multigene family of *Listeria monocytogenes* EGD. *Infect Immun.*, 1997, 65(5), 1615-1625.

Duche, O; Tremoulet, F; Glaser, P; Labadie, J. Salt stress proteins induced in *Listeria monocytogenes. Appl. Environ. Microbiol.,* 2002, 68 (4), 1491–1498.

Durack, J; Ross, T; Bowman, JP. Characterisation of the transcriptomes of genetically diverse *Listeria monocytogenes* exposed to hyperosmotic and low temperature conditions reveal global stress-adaptation mechanisms. *PLoS ONE.,* 2013, 8(9), e73603. doi:10.1371/journal.pone.0073603

Dutta, V; Elhanafi, D; Kathariou, S. Conservation and distribution of the benzalkonium chloride resistance cassette bcrABC in *Listeria monocytogenes. Appl. Environ. Microbiol.,* 2013, 79(19), 6067–74.

Elhanafi, D; Dutta, V; Kathariou, S. Genetic characterization of plasmid-associated benzalkonium chloride resistance determinants in a *Listeria monocytogenes* strain from the 1998–1999 outbreak. *App. Environ. Microbiol.,* 2010,76(24), 8231–8.

Fallah, AA; Saei-Dehkordi, SS; Rahnama, M; Tahmasby, H; Mahzounieh, M. Prevalence and antimicrobial resistance patterns of *Listeria* species isolated from poultry products marketed in Iran. *Food Control*, 2012, 28, 327-332.

Farber, JM; Addison, CJ. RAPD typing for distinguishing species and strains in the genus *Listeria. J. Appl. Bacteriol.,* 1994, 77, 242–250.

Farber, JM; Peterkin, PI. *Listeria monocytogenes*, a food-borne pathogen, *Microbiol. Rev.*, 1991, 55, 476.

Farlow, J; Smith, KL; Wong, J; Abrams, M; Lytle, M; Keim, P. *Francisella tularensis* strain typing using multiple-locus, variable-number tandem repeat analysis. *J. Clin. Microbiol.*, 2001, 39, 3186-3192.

Feil, EJ; Cooper, JE; Grundmann, H; Robinson, DA; Enright, MC; Berendt, T; Peacock, SJ; Smith, JM; Murphy, M; Spratt, BG; Moore, CE; Day, NPJ. How clonal is *Staphylococcus aureus*? *J. Bacteriol.*, 2003, 185, 3307-3316.

Fouet, A; Namy, O; Lambert, G. Characterization of the operon encoding the alternative _B factor from *Bacillus anthracis* and its role in virulence. *J. Bacteriol.*, 2000, 182, 5036–5045.

Franciosa, G; Tartaro, S; Wedell-Neergaard, C; Aureli, P. Characterization of *Listeria monocytogenes* strains involved in invasive and noninvasive listeriosis outbreaks by PCR-based fingerprinting techniques. *Appl. Environ. Microbiol.,* 2001, 67, 1793–1799.

Francis, GA; Thomas, C; O'Beirne, D. The microbiological safety of minimally processed vegetables. *Int. J. Food Sci. Tech.,* 1999, 34, 1-22.

Furrer, B; Candrian, U; Hoefelein, Ch; Luethy, J. Detection and identification of *Listeria monocytogenes* in cooked sausage products and in milk by in vitro amplification of haemolysin gene fragments. *J. Appl. Bacteriol.*, 1991, 70, 372–379.

Gahan, CG; O'Mahony, J; Hill, C. Characterization of the *groESL* operon in *Listeria monocytogenes*: utilization of two reporter systems (*gfp* and *hly*) for evaluating in vivo expression. *Infect. Immun.,* 2001, 69, 3924–3932.

Gaillot, O; Pellegrini, E; Bregenholt, S; Nair, S; Berche, P. The ClpP serine protease is essential for the intracellular parasitism and virulence of *Listeria monocytogenes. Mol. Microbiol.,* 2000, 35(6), 1286-1294.

Gething, MJH; Sambrook, JF. Protein folding in the cell. *Nature,* 1992, 355, 33-45.

Giovannacci, I; Ragimbeau, C; Queguiner, S; Salvat, G; Vendeuvre, JL; Carlier, V; Ermel. G. *Listeria monocytogenes* in pork slaughtering and cutting plants: use of RAPD, PFGE and PCR–REA for tracing and molecular epidemiology. *Int. J. Food Microbiol.*, 1999, 53, 127–140.

Gorski, L; Walker, S; Liang, AS; Nguyen, KM; Govoni, J; Carychao, D; Cooley, MB; Mandrell, RE. Comparison of subtypes of *Listeria monocytogenes* isolates from naturally contaminated watershed samples with and without a selective secondary enrichment. *PLoS ONE.*, 2014, 9(3): e92467

Graumann, PL; Marahiel, MA. A superfamily of proteins that contain the cold-shock domain. *Trends Biochem. Sci.,* 1998, 23, 286–290.

Grundling, A; Gonzalez, MD; Higgins, DE. Requirement of the *Listeria monocytogenes* broad-range phospholipase PC-PLC during infection of human epithelial cells. *J. Bacteriol.,* 2003, 185, 6295–6307.

Guariglia-Oropeza, V; Orsi, RH; Yu, H; Boor, KJ; Wiedmann, M; Guldimann, C. Regulatory network features in *Listeria monocytogenes*—changing the way we talk. *Front. Cell. Infect. Microbiol.,* 2014, 4, 14, 1.

Haase, JK; Didelot, X; Lecuit, M; Korkeala H. The ubiquitous nature of *Listeria monocytogenes* clones: a large-scale multilocus sequence typing study. *Environ. Microbiol.,* 2014, 16(2), 405–416.

Hadjilouka, A; Andritsos, ND; Paramithiotis, S; Mataragas, M; Drosinos, EH. *Listeria monocytogenes* serotype prevalence and biodiversity in diverse food products. *J. Food Prot.,* 2014, 77 (12), 2115–2120.

Hamon, MA; Ribet, D; Stavru F; Cossart P. Listeriolysin O: the Swiss army knife of *Listeria. Trends Microbiol.*, 2012, 20, 360-368.

Han, Y; Linton, RH; Nielsen, SS; Nelson, PE. Reduction of *Listeria monocytogenes* on green peppers (*Capsicum annuum* L.) by gaseous and aqueous chlorine dioxide and water washing and its growth at 7 degrees C. *J. Food Prot.,* 2001, 64, 1730–1738.

Hanawa, T; Yamanishi, S; Murayama, S; Yamamoto, T; Kamiya, S. Participation of DnaK in expression of genes involved in virulence of *Listeria monocytogenes. FEMS Microbiol. Lett.,* 2002, 214, 69-75.

Hanawa, T; Fukuda, M; Kawakami, H; Hirano, H; Kamiya, S; Yamamoto; T.The *Listeria monocytogenes* DnaK chaperone is required for stress tolerance and efficient phagocytosis with macrophages. *Cell Stress Chaperones*, 1999, 4, 118–128.

Hartl, FU. Molecular chaperones in cellular protein folding. *Nature*, 1996, 381, 571-580.

Hess, J; Gentschev, I; Szalay, G; Ladel, C; Bubert, A; Goebel, W; Kaufmann, SH. *Listeria monocytogenes* p60 supports host cell invasion by and in vivo survival of attenuated *Salmonella typhimurium. Infect. Immun.,* 1995, 63, 2047-2053.

Hoe, N; Nakashima, K; Grigsby, D; Pan, X; Dou, SJ; Naidich, S; Garcia, M; Kahn, E; Bergmire-Sweat, D; Musser, JM. Rapid molecular genetic subtyping of serotype M1 group A *Streptococcus* strains. *Emerg. Infect. Dis.*, 1999, 43, 254-263.

Hopkins, KL; Maguire, C; Best, E; Liebana, E; Threlfall, EJ. Stability of multiple-locus variable-number tandem repeats in *Salmonella enterica* serovar Typhimurium. *J. Clin. Microbiol.*, 2007, 45, 3058-3061.

Horvath, P; Barrangou, R. CRISPR/Cas, the immune system of bacteria and archaea. *Science,* 2010, 327, 167-170.

Horvath, P; Coute-Monvoisin, AC; Romero, DA; Boyaval, P; Fremaux, C; Barrangou, R. Comparative analysis of CRISPR loci in lactic acid bacteria genomes. *Int. J. Food Microbiol.,* 2009, 131, 62-70.

Hunter, PR; Gaston, MA. Numerical index of the discriminatory ability of typing systems: an application of Simpson's index of diversity. *J. Clin. Microbiol.,* 1988, 26, 2465–2466.

International Organization for Standardization. 1996. Microbiology of food and animal feeding stuff—horizontal method for the detection and enumeration of *Listeria monocytogenes*. Part 1. Detection method. International Organization for Standardization, Geneva.

Iredell, J; Blanckenberg, D; Arvand, M; Grauling, S; Feil, EJ; Birtles, RJ. Characterization of the natural population of *Bartonella henselae* by multilocus sequence typing. *J. Clin. Microbiol.*, 2003, 41, 5071-5079.

Jacquet, C; Doumith, M; Gordon, JI; Martin, PM; Cossart, P; Lecuit, M. A molecular marker for evaluating the pathogenic potential of foodborne *Listeria monocytogenes. J. Infect. Dis*., 2004, 189, 2094–2100.

Jamali, H; Radmehr B. Frequency, virulence genes and antimicrobial resistance of *Listeria* spp. isolated from bovine clinical mastitis. *Vet. J.*, 2013, 198, 541–542.

Jami, M; Ghanbari, M; Zunabovic, M; Domig, KJ; Kneifel, W. *Listeria monocytogenes* in Aquatic Food Products—A Review. *Compr. Rev. Food Sci. F.,* 2014, 13, 798-813.

Jaradat, ZW; Schutze, GE; Bhunia AK. Genetic homogeneity among *Listeria monocytogenes* strains from infected patients and meat products from two geographic locations determined by phenotyping, ribotyping and PCR analysis of virulence genes, *Int. J. Food Microbiol.,* 2002, 76, 1-10.

Jeon, SM; Lim, NR; Kwon, SJ; Shim, TS; Park, MS; Kim, BJ; Kim, SH. Analysis of species and intra-species associations between the *Mycobacterium abscessus* complex strains using pulsed-field gel electrophoresis (PFGE) and multi-locus sequence typing (MLST). *J. Microbiol. Meth.*, 2014, 104, 19-25.

Jersek, B; Gilot, P, Gubina, M; Klun, N; Mehle, J; Tcherneva, E; Rijpens, N; Herman, L. Typing of *Listeria monocytogenes* strains by repetitive element sequence–based PCR. *J. Clin. Microbiol.,* 1999, 37, 103–109.

Jia, Y; Nightingale, KK; Boor, KJ; Ho, A; Wiedmann, M; McGann, P. Distribution of internalin gene profiles of *Listeria monocytogenes* isolates from different sources associated with phylogenetic lineages. *Foodborne Pathog. Dis.,* 2007, 4, 222–232.

Johansson, A; Farlow, J; Larsson, P; Dukerich, M; Chambers, E; Bystrom, M; Fox, J; Chu, M; Forsman, M; Sjostedt, A; Keim, P. Worldwide genetic relationships among *Francisella tularensis* isolates determined by multiple-locus variable-number tandem repeat analysis. *J. Bacteriol.*, 2004, 186, 5808-5818.

Kallipolitis, BH; Ingmer, H. *Listeria monocytogenes* response regulators important for stress tolerance and pathogenesis. *FEMS Microbiol. Lett.*, 2001, 204, 111.

Kathariou, S; Metz, P; Hof, H; Goebel, W. Tn916-induced mutations in the hemolysin determinant affecting virulence of *Listeria monocytogenes*. *J. Bacteriol.,* 1987, 169, 1291–1297.

Kaur, S; Malik, SVS; Vaidya; VM, Barbuddhe, SB. *Listeria monocytogenes* in spontaneous abortions in humans and its detection by multiplex PCR. J. Appl. Microbiol., 2007, 103, 1889–1896.

Keim, P; Price, LB; Klevytska, AM; Smith, KL; Schupp, JM; Okinaka, R; Jackson, PJ; Hugh-Jones, ME. Multiple-locus variable-number tandem repeat analysis reveals genetic relationships within *Bacillus anthracis*. *J. Bacteriol*., 2000, 182, 2928-2936.

Keto-Timonen, RO; Autio, TJ; Korkeala, HJ. An improved amplified fragment length polymorphism (AFLP) protocol for discrimination of *Listeria* isolates. *Syst. Appl. Microbiol.,* 2003, 26, 236–244.

Khen, BK; Lynch, OA; Carroll1, J; McDowell, DA; Duffy, G. Occurrence, antibiotic resistance and molecular characterization of *Listeria monocytogenes* in the beef chain in the Republic of Ireland. *Zoonoses Public Hlth.,* 2015, 62, 11–17.

King, SJ; Leigh, JA; Heath, PJ; Luque, I; Tarradas, C; Dowson, CG; Whatmore, AM. Development of a multilocus sequence typing scheme for the pig pathogen *Streptococcus suis*: identification of virulent clones and potential capsular serotype exchange. *J. Clin. Microbiol*., 2002, 40, 3671-3680.

Klevytska, AM; Price, LB; Schupp, JM; Worsham, PL; Wong, J; Keim, P. Identification and characterization of variable-number tandem repeats in the *Yersinia pestis* genome. *J. Clin. Microbiol*., 2001, 39, 3179-3185.

Kotetishvili, M; Stine, OC; Chen, Y; Kreger, A; Sulakvelidze, A; Sozhamannan, S; Morris Jr, JG. Multilocus sequence typing has better discriminatory ability for typing *Vibrio cholerae* than does pulsed-field gel electrophoresis and provides a measure of phylogenetic relatedness. *J. Clin. Microbiol*., 2003, 41, 2191-2196.

Kotetishvili, M; Stine, OC; Kreger, A; Morris Jr, JG; Sulakvelidze, A. Multilocus Sequence typing for characterization of clinical and environmental *Salmonella* strains. *J. Clin. Microbiol*., 2002, 40, 1626-1635.

Kramarenko, T; Roasto, M; Meremäe, K; Kuningas, M; Põltsama, P; Elias, T. *Listeria monocytogenes* prevalence and serotype diversity in various foods. *Food Control*, 2013, 30, 24-29.

Laursen, MF; Bahl, MI; Licht, TR; Gram, L; Knudsen, GM. A single exposure to a sublethal pediocin concentration initiates a resistance-associated temporal cell envelope and general stress response in *Listeria monocytogenes. Environ. Microbiol.,* 2015, 17, 1134–1151.

Lawrence, LM; Harvey, J; Gilmour, A. Development of a random amplification of polymorphic DNA typing method for *Listeria monocytogenes*. *Appl. Environ. Microbiol.,* 1993, 59, 3117-3119.

Li, X; Huang, B; Eglezos, S; Graham, T; Blair, B; Bates, J. Identification of an optimized panel of variable number tandem-repeat (VNTR) loci for *Listeria monocytogenes* typing. *Diagn. Microbiol. Infect. Dis.,* 2013, 75, 203–206.

Liang, SY; Watanabe, H; Terajima, J; Li, CC; Liao, JC; Tung, SK; Chiou, CS. Multilocus variable-number tandem repeat analysis for molecular typing of *Shigella sonnei*. *J. Clin. Microbiol*., 2007, 45, 3574-3580.

Lin, CM; Fernando, SY; Wei, CI. Occurrence of *Listeria monocytogenes*, *Salmonella* spp., *Escherichia coli* and *E. coli* O157:H7 in vegetable salads. *Food Control*, 1996, 7, 135-40.

Linder, P; Jankowsky, E. From unwinding to clamping—the DEAD box RNA helicase family. *Nat. Rev. Mol. Cell Biol.,* 2011, 12, 505–516.

Lindstedt, BA; Tham, W; Danielsson-Tham, ML; Vardund, T; Helmersson, S; Kapperud, G. Multiple-locus variable-number tandem-repeats analysis of *Listeria monocytogenes* using multicolour capillary electrophoresis and comparison with pulsed-field gel electrophoresis typing. *J. Microbiol. Meth.,* 2008, 72, 141–148.

Lindstedt, BA; Heir, E; Gjernes, E; Kapperud G. DNA Fingerprinting of *Salmonella enterica* subsp. *enterica* serovar Typhimurium with emphasis on phage type DT104 based on variable number of tandem repeat loci. *J. Clin. Microbiol.*, 2003a, 41, 1469-1479.

Lindstedt, BA; Heir, E; Gjernes, E; Vardund, T; Kapperud, G. DNA fingerprinting of Shiga-toxin producing *Escherichia coli* O157 based on multiple-locus variable-number tandem-repeats analysis (MLVA). *Ann. Clin. Microbiol. Antimicrob*., 2003b, 2, 12.

Liu, X; Basu, U; Miller, P; McMullen, LM. Stress response and adaptation of *Listeria monocytogenes* 08-5923 exposed to a sublethal dose of carnocyclin. *Appl. Environ. Microbiol.,* 2014, 80 (13), 3835–3841.

Liu, D; Lawrence, ML; Austin, FW; Ainsworth, AJ. A multiplex PCR for species- and virulence-specific determination of *Listeria monocytogenes. J. Microbiol. Meth.*, 2007, 71, 133–140.

Liu, Y; Ream, A. Gene expression profiling of *Listeria monocytogenes* strain F2365 during growth in ultrahigh-temperature-processed skim milk. *Appl. Environ. Microbiol.,* 2008, 74 (22), 6859–6866.

Lomonaco, S; Patti, R; Knabel, SJ; Civera, T. Detection of virulence-associated genes and epidemic clone markers in *Listeria monocytogenes* isolates from PDO Gorgonzola cheese. *Int. J. Food Microbiol*., 2012, 160, 76–79.

López-Gálvez, F; Allende, A; Truchado, P; Martínez-Sánchez, A; Tudela, JA; Selma, MV; Gil, MI. Suitability of aqueous chlorine dioxide versus sodium hypochlorite as an effective sanitizer for preserving quality of fresh-cut lettuce while avoiding by-product formation. *Postharvest. Biol. Tec*., 2010, 55, 53–60.

Louwen, R; Staals, RHJ; Endtz, HP; van Baarlen, P; van der Oost, J. The Role of CRISPR-Cas systems in virulence of pathogenic bacteria. *Microbiol. Mol. Biol. R.,* 2014, 78 (1), 74–88.

Lunestad, BT; Truong, TTT; Lindstedt, BA. A multiple-locus variable-number tandem repeat analysis (MLVA) of *Listeria monocytogenes* isolated from Norwegian salmon-processing factories and from listeriosis patients. *Epidemiol. Infect.,* 2013, 141, 2101–2110.

Mahmoud, BS; Vaidya, NA; Corvalan, CM; Linton, RH. Inactivation kinetics of inoculated *Escherichia coli* O157:H7, *Listeria monocytogenes* and *Salmonella* Poona on whole cantaloupe by chlorine dioxide gas. *Food Microbiol.,* 2008, 25, 857–865.

Maiden, MCJ; Bygraves, JA; Feil, E; Morelli, G; Russell, JE; Urwin, R; Zhang, Q; Zhou, J; Zurth, K; Caugant, DA; Feavers, IM; Achtman, M; Spratt, BG. Multilocus sequence typing: A portable approach to the identification of clones within populations of pathogenic microorganisms. *Proc. Natl. Acad. Sci. USA*, 1988, 95, 3140-3145.

Mammina, C; Parisi, A; Guaita, A; Aleo, A; Bonura, C; Nastasi, A; Pontello, M. Enhanced surveillance of invasive listeriosis in the Lombardy region, Italy, in the years 2006-2010 reveals major clones and an increase in serotype 1/2a. *BMC Infect Dis*., 2013, 13, 152.

Markkula, A; Mattila, M; Lindström, M; Korkeala, H. Genes encoding putative DEAD-box RNA helicases in *Listeria monocytogenes* EGD-e are needed for growth and motility at 3°C. *Environ. Microbiol.,* 2012, 14, 2223–2232.

Markkula, A; Lindström, M; Johansson, P; Björkroth, J; Korkeala, H. Roles of four putative DEAD-Box RNA helicase genes in growth of *Listeria monocytogenes* EGD-e under heat, pH, osmotic, ethanol, and oxidative stress conditions. *Appl. Environ. Microbiol.,* 2012a, 78(19), 6875–6882.

Marraffini, LA; Sontheimer, EJ. CRISPR interference: RNA-directed adaptive immunity in bacteria and archaea. *Nat. Rev. Genet.,* 2010, 11, 181-190.

Martín, B; Perich, A; Gomez, D; Yangüela, J; Rodríguez, A; Garriga, M; Aymerich, T. Diversity and distribution of *Listeria monocytogenes* in meat processing plants. *Food Microbiol.,* 2014, 44 119-127.

Martin-Visscher, LA; van Belkum, MJ; Garneau-Tsodikova, S; Whittal, RM; Zheng, J; McMullen, LM; Vederas, JC. Isolation and characterization of carnocyclin A, a novel circular bacteriocin produced by *Carnobacterium maltaromaticum* UAL307. *Appl. Environ. Microbiol.,* 2008,74, 4756–4763.

Mataragas, M; Rovetto, F; Bellio, A; Alessandria, V; Rantsiou, K; Decastelli, L; Cocolin, L. Differential gene expression profiling of *Listeria monocytogenes* in Cacciatore and Felino salami to reveal potential stress resistance biomarkers. *Food Microbiol.,* 2015, 46, 408-417.

McDonnell, G; Russell, AD. Antiseptics and disinfectants: activity, action, and resistance. *Clin. Microbiol. Rev.,* 1999, 12, 147–179.

McTavish, SM; Pope, CE; Nicol, C; Campbell, D; French, N; Carter, PE. Multilocus sequence typing of *Campylobacter jejuni*, and the correlation between clonal complex and pulsed-field gel electrophoresis macrorestriction profile. *FEMS Microbiol. Lett.*, 2009, 298, 149-156.

Milecka, D; Samluk, A; Wasiak, K; Krawczyk⍰Balska, A. An essential role of a ferritin⍰like protein in acid stress tolerance of *Listeria monocytogenes. Arch. Microbiol.,* 2015, 197, 347–351.

Miya, S; Kimura, B; Sato, M; Takahashi, H; Ishikawa, T; Suda, T; Takakura, C; Fujii, T; Wiedmann, M. Development of a multilocus variable-number of tandem repeat typing method for *Listeria monocytogenes* serotype 4b strains. *Int. J. Food Microbiol.,* 2008, 124, 239–249.

Miya, S; Takahashi, H; Kamimura, C; Nakagawa, M; Kuda, T; Kimura, B. Highly discriminatory typing method for *Listeria monocytogenes* using polymorphic tandem repeat regions. *J. Microbiol. Meth.,* 2012, 90, 285–291.

Montero, D; Bodero, M; Riveros, G; Lapierre, L; Gaggero, A;Vidal, RM; Vidal, M. Molecular epidemiology and genetic diversity of *Listeria monocytogenes* isolates from a wide variety of ready-to-eat foods and their relationship to clinical strains from listeriosis outbreaks in Chile. *Front. Microbiol.,* 2015, 6, 384.

Moreno, LZ; Paixão, R; de Gobbi, DDS; Raimundo, DC; Ferreira, TSP; Moreno, AM; Hofer, E; dos Reis, CMF; Matté, GR; Matté, MH. Phenotypic and genotypic characterization of a typical *Listeria monocytogenes* and *Listeria innocua* isolated from swine slaughterhouses and meat markets. BioMed Research International. Hindawi Publishing Corporation. Volume 2014, Article ID 742032, 12 pages. http://dx.doi.org/ 10.1155/2014/742032.

Mujahid, S; Orsi, RH; Boor, KJ; Wiedmann, M. Protein level identification of the *Listeria monocytogenes* Sigma H, Sigma L, and Sigma C regulons. *BMC Microbiol.,* 2013, 13, 156.

Murphy, M; Corcoran, D; Buckley, JF; O'Mahony, M; Whyte, P; Fanning, S. Development and application of Multiple-Locus Variable Number of tandem repeat Analysis (MLVA) to subtype a collection of *Listeria monocytogenes. Int. J. Food Microbiol.,* 2007, 115, 187–194.

Nair, S; Derre, I; Msadek, T; Gaillot, O; Berche, P. CtsR controls class III heat shock gene expression in the human pathogen *Listeria monocytogenes. Mol. Microbiol.,* 2000b, 35(4), 800-811.

Nair, S; Milohanic, E; Berche, P. ClpC ATPase is required for cell adhesion and invasion of *Listeria monocytogenes. Infect. Immun.,* 2000a, 68 (12), 7061–7068.

Nair, S; Frehel, C; Nguyen, L; Escuyer, V; Berche P. ClpE, a novel member of the HSP100 family, is involved in cell division and virulence of *Listeria monocytogenes. Mol. Microbiol.,* 1999, 31, 185–196.

Nelson, KE; Fouts, DE; Mongodin, EF; Ravel, J; DeBoy, RT; Kolonay, JF, Rasko, DA; Angiuoli, SV; Gill, SR; Paulsen IT; Peterson, J; White, O; Nelson, WC; Nierman, W; Beanan, MJ; Brinkac, LM; Daugherty, SC; Dodson, RJ; Durkin, AS; Madupu, R; Haft, DH; Selengut, J; Van Aken, S; Khouri, H; Fedorova, N; Forberger, H; Tran, B; Kathariou, S. Wonderling, LD; Uhlich, GA; Bayles, DO;. Luchansky, JB; Fraser, CM. Whole genome comparisons of serotype 4b and 1/2a strains of the food-borne pathogen *Listeria monocytogenes* reveal new insights into the core genome components of this species. *Nucleic Acids Res.,* 2004, 32, 2386–2395.

Noller, AC; McEllistrem, MC; Pacheco, AG; Boxrud, DJ; Harrison, LH. Multilocus variable-number tandem repeat analysis distinguishes outbreak and sporadic *Escherichia coli* O157:H7 isolates. *J. Clin. Microbiol.*, 2003, 41, 5389-5397.

Notermans, SHW; Dufrenne, J; Leimeister-Wachter, M; Domann, E; Chakraborty, T. Phosphatidylinositol-specific phospholipase C activity as a marker to distinguish between pathogenic and non-pathogenic *Listeria* species. *Appl. Environ. Microbiol.*, 1991, 57, 2666–2670.

Obaidat, MM; Salman, AEB; Lafi, SQ; Al-Abboodi, AR. Characterization of *Listeria monocytogenes* from three countries and antibiotic resistance differences among countries and *Listeria monocytogenes* serogroups. *Lett. Appl. Microbiol.,* 2015, 60, 609-614.

Okada, Y; Makino, S; Tobe, T; Okada, N; Yamazaki, S. Cloning of rel from *Listeria monocytogenes* as an osmotolerance involvement gene. *Appl, Environ, Microbiol.,* 2002, 68, 1541–1547.

Olesen, I; Thorsen, L; Jespersen, L. Relative transcription of *Listeria monocytogenes* virulence genes in liver pâtés with varying NaCl content. *Int. J. Food Microbiol.*, 2010, 141, S60–S68.

Oliver, HF; Orsi, RH; Wiedmann, M; Boor, KJ. *Listeria monocytogenes* sigma B has a small core regulon and a conserved role in virulence but makes differential contributions to stress tolerance across a diverse collection of strains. *Appl. Environ. Microbiol.,* 2010, 76 (13), 4216–4232.

Onteniente, L; Brisse, S; Tassios, PT; Vergnaud, G. Evaluation of the polymorphisms associated with tandem repeats for *Pseudomonas aeruginosa* strain typing. *J. Clin. Microbiol.*, 2003, 41, 4991-4997.

Orsi, RH; den Bakker, HC; Wiedmann. M. *Listeria monocytogenes* lineages: Genomics, evolution, ecology, and phenotypic characteristics. *Int. J. Med. Microbiol.*, 2011, 301, 79–96.

Overduin, P; Schouls, L; Roholl, P; van der Zanden, A; Mahmmod, N; Herrewegh, A; van Soolingen, D. Use of multilocus variable-number tandem-repeat analysis for typing *Mycobacterium avium* subsp. *paratuberculosis*. *J. Clin. Microbiol.*, 2004, 42, 5022-5028.

Paramithiotis, S; Hadjilouka, A; Drosinos, EH.. Listeria pathogenicity island 1. Structure and function. In: E.C. Hambrick (Ed.) *Listeria monocytogenes*: Food Sources, Prevalence and Management Strategies. Nova Publishers, 2014, pp. 265-282.

Paramithiotis, S; Kagkli, DM; Blana, VA; Nychas, GJE; Drosinos, EH. Identification and characterization of *Enterococcus* spp. in Greek spontaneous sausage fermentation. *J. Food Prot.*, 2008, 71, 1244-1247.

Parisi, A; Latorre, L; Normanno, G; Miccolupo, A; Fraccalvieri, R; Lorusso, V; Santagada, G. Amplified fragment length polymorphism and multi-locus sequence typing for high-resolution genotyping of *Listeria monocytogenes* from foods and the environment. *Food Microbiol.*, 2010, 27, 101–108.

Paziak-Domanska, B; Bogulawska, E; Wiekowska-Szakiel, M; Kotlowski, R; Rozalska, B; Chmiela, M; Kur, J; Dabrowski, W; Rudnicka, W. Evaluation of the API test, phosphatidylinositol-specific phospholipase C activity and PCR method in identification of *Listeria monocytogenes* in meat foods. *FEMS Microbiol. Lett.*, 1999, 171, 209–214.

Pettengill, JB; Timme, RE; Barrangou, R; Toro, M; Allard, MW; Strain, E; Musser SM; Brown, EW. The evolutionary history and diagnostic utility of the CRISPR-Cas system within *Salmonella enterica* ssp. *enterica*. *PeerJ.*, 2014, 2, e340.

Phan-Thanh, L; Gormon, T. Analysis of heat and cold shock proteins in *Listeria* by two-dimensional electrophoresis. *Electrophoresis,* 1995, 16, 444–450.

Pleitner, AM; Trinetta, V; Morgan, MT; Linton, RL; Oliver, HF. Transcriptional and phenotypic responses of *Listeria monocytogenes* to chlorine dioxide. *Appl. Environ. Microbiol.*, 2014, 80 (9), 2951–2963.

Pourcel, C; Salvignol, G; Vergnaud, G. CRISPR elements in *Yersinia pestis* acquire new repeats by preferential uptake of bacteriophage DNA, and provide additional tools for evolutionary studies. *Microbiol.*, 2005, 151, 653-663.

Raffelsbauer, D; Bubert, A; Engelbrecht, F; Scheinpflug, J; Simm, A; Hess, J; Kaufmann, SH; Goebel, W. The gene cluster inlC2DE of *Listeria monocytogenes* contains additional new internalin genes and is important for virulence in mice. *Mol. Gen. Genet.*, 1998, 260, 144–158.

Ragon, M; Wirth, T; Hollandt, F; Lavenir, R; Lecuit, M; Le, MA; Brisse, S. A new perspective on *Listeria monocytogenes* evolution. *PLoS Pathog.*, 2008, 4, e1000146.

Rawool, DB; Malik, SVS; Shakuntala, I; Sahare, AM; Barbuddhe, SB. Detection of multiple virulence-associated genes in *Listeria monocytogenes* isolated from bovine mastitis cases. *Int. J. Food Microbiol.*, 2007, 113, 201-207.

Revazishvili, T; Kotetishvili, M; Stine, OC; Kreger, AS; Morris, JrJG.; Sulakvelidze, A. Comparative analysis of multilocus sequence typing and pulsed-field gel electrophoresis for characterizing *Listeria monocytogenes* strains isolated from environmental and clinical sources. *J. Clin. Microbiol.*, 2004, 42, 276–285.

Ripio, MT; Vazquez-Boland, JA; Vega, Y; Nair, S; Berche, P. Evidence for expressional crosstalk between the central virulence regulator PrfA and the stress response mediator ClpC in *Listeria monocytogenes. FEMS Microbiol. Lett.*, 1998, 158, 45-50.

Roche, SM; Velge, P; Liu, D. Virulence determination. In: D. Liu (Ed.) Hanbook of *Listeria monocytogenes*. CRC Press, 2008, pp. 241-270.

Romanova, NA; Wolffs, PF; Brovko, LY; Griffiths, MW. Role of efflux pumps in adaptation and resistance of *Listeria monocytogenes* to benzalkonium chloride. *Appl. Environ. Microbiol.,* 2006, 72(5), 3498–503.

Ross, IL; Parkinson, IH; Heuzenroeder, MW. The use of MAPLT and MLVA analyses of phenotypically closely related isolates of *Salmonella enterica* serovar Typhimurium. *Int. J. Med. Microbiol.*, 2009, 299, 37-41.

Rouquette, C; de Chastellier, C; Nair, S; Berche P. The ClpC ATPase of *Listeria monocytogenes* is a general stress protein required for virulence and promoting early bacterial escape from the phagosome of macrophages. *Mol. Microbiol.,* 1998. 27, 1235–1245.

Rouquette, C; Ripio, MT; Pellegrini, EJ; Bolla, M; Tascon, RI; Vazquez-Boland, JA;. Berche, P. Identification of a ClpC ATPase required for stress tolerance and in vivo survival of *Listeria monocytogenes. Mol. Microbiol.,* 1996, 21, 977–987.

Ryu, JH; Beuchat, LR. Biofilm formation and sporulation by *Bacillus cereus* on a stainless steel surface and subsequent resistance of vegetative cells and spores to chlorine, chlorine dioxide, and a peroxyacetic acid based sanitizer. *J. Food Prot.,* 2005, 68, 2614–2622.

Sabet, C; Lecuit, M; Cabanes, D; Cossart, P; Bierne H. LPXTG protein InlJ, a newly identified internalin involved in *Listeria monocytogenes* virulence, *Infect. Immun.,* 2005, 73, 6912–6922.

Sado, PN; Jinneman, KC; Husby, GJ; Sorg, SM; Omiecinski, CJ. Identification of *Listeria monocytogenes* from unpasteurized apple juice using rapid test kits. *J. Food Prot.,* 1998, 61, 1199-202.

Salcedo, C; Arreaza, L; Alcala, B; de la Fuente, L; Vazquez, JA.. Development of a multilocus sequence typing method for analysis of *Listeria monocytogenes* clones. *J. Clin. Microbiol.*, 2003, 41, 757–762.

Saleh-Lakha, S; Allen, VG; Li, J; Pagotto, F; Odumeru, J; Taboada, E; Lombos, M; Tabing, KC; Blais, B; Ogunremi, D; Downing, G; Lee, S; Gao, A; Nadon, C; Chen, S. Subtyping of a large collection of historical *Listeria monocytogenes* strains from Ontario, Canada, by an improved Multilocus Variable- Number Tandem-Repeat Analysis (MLVA). Appl. Environ. Microbiol., 2013, 79 (20), 6472–6480.

Schmid, B; Klumpp, J; Raimann, E; Loessner, MJ; Stephan, R; Tasara, T. Role of cold shock proteins in growth of *Listeria monocytogenes* under cold and osmotic stress conditions. *Appl. Environ. Microbiol.,* 2009, 75 (6), 1621–1627.

Schouls, LM; Reulen, S; Duim, B; Wagenaar, JA; Willems, RJ; Dingle, KE; Colles, FM; Van Embden, JD. Comparative genotyping of *Campylobacter jejuni* by amplified fragment length polymorphism, multilocus sequence typing, and short repeat sequencing: strain diversity, host range, and recombination. *J. Clin. Microbiol.,* 2003, 41, 15–26.

Schouls, LM; van der Ende, A; van de Pol, I; Schot, C; Spanjaard, L; Vauterin, P; Wilderbeek, D; Witteveen, S. Increase in genetic diversity of *Haemophilus influenzae* serotype b (Hib) strains after introduction of Hib vaccination in The Netherlands. *J. Clin. Microbiol.*, 2005, 43, 2741-2749.

Schouls, LM; van der Heide, HG; Vauterin, L; Vauterin, P; Mooi, FR. Multiple-locus variable-number tandem repeat analysis of Dutch *Bordetella pertussis* strains reveals rapid genetic changes with clonal expansion during the late 1990s. *J. Bacteriol.*, 2004, 186, 5496-5505.

Sesto, N; Touchon, M; Andrade, J M; Kondo, J; Rocha, EPC; Arraiano, CM; Archambaud, C; Westhof, E; Romby, P; Cossart, P. A PNPase Dependent CRISPR System in *Listeria. PLoS Genet.*, 2014, 10 (1): e1004065.

Shen, J; Rumpb, L; Zhang, Y; Chen, Y; Wang, X; Meng, J. Molecular subtyping and virulence gene analysis of *Listeria monocytogenes* isolates from food. *Food Microbiol.*, 2013, 35, 58-64.

Sleator, RD; Hill, C. A novel role for the LisRK two-component regulatory system in listerial osmotolerance. *Clin. Microbiol. Infect.*, 2005, 11, 599–601.

Sleator, RD; Gahan, CG; Abee, T; Hill, C. Identification and disruption of BetL, a secondary glycine betaine transport system linked to the salt tolerance of *Listeria monocytogenes* LO28. *Appl. Environ. Microbiol.*, 1999, 65, 2078–2083.

Sleator, RD; Hill, C. A novel role for the LisRK two-component regulatory system in listerial osmotolerance, *Clin. Micobiol. Infect.*, 2005, 11, 599.

Sleator, RD; Gahan, CGM; Hill, C. Identification and disruption of the *proBA* locus in *Listeria monocytogenes*: Role of proline biosythesis in salt tolerance and murine infection, *Appl. Environ. Microbiol.*, 2001a, 67, 2571.

Sleator, RD; Gahan, CGM; Hill, C. Mutations in the listerial *proB* gene leading to proline overproduction: Effects on salt tolerance and murine infection, *Appl. Environ. Microbiol.*, 2001b, 67, 4560.

Sommerville, J; Ladomery, M. Masking of mRNA by Y-box proteins. *FASEB J.*, 1996, 10, 435–443.

Soni, DK; Singh, M; Singh, DV; Dubey, SK. Virulence and genotypic characterization of *Listeria monocytogenes* isolated from vegetable and soil samples. *BMC Microbiol.*, 2014, 14, 241.

Stack, HM; Hill, C; Gahan, CGM. 2008. Stress Responses. In: D. Liu (Ed.) Hanbook of *Listeria monocytogenes*. CRC Press, 2008, pp. 61-98.

Stack, HM; Sleator, RD; Bowers, M; Hill, C; Gahan, CGM. Role for HtrA in stress induction and virulence potential in *Listeria monocytogenes. Appl. Environ. Microbiol.,* 2005, 71 (8), 4241–4247.

Suarez, M; Vazquez-Boland, JA. The bacterial actin nucleter protein ActA is involved in epithelial cell invasion by *Listeria monocytogenes*. 2001. www.ncbi.nlm.nih.gov/pubmed [Accession No. AF103807].

Sukhnanand, S; Alcaine, S; Warnick, LD; Su, WL; Hof, J; Pat, M; Craver, J; McDonough, P; Boor, KJ; Wiedmann M. DNA sequence-based subtyping and evolutionary analysis of selected *Salmonella enterica* serotypes. *J. Clin. Microbiol.*, 2005, 43, 3688-3698.

Svraka, S; Toman, R; Skultety, L; Slaba, K; Homan, WL. Establishment of a genotyping scheme for *Coxiella burnetii. FEMS Microbiol. Lett.*, 2006, 254, 268-274.

Swaminathan, B; Barrett, TJ; Hunter, SB; Tauxe, RV. PulseNet: the molecular subtyping network for foodborne bacterial disease surveillance, United States. *Emerg. Infect. Dis.*, 2001, 7, 382-389.

Tamburro, M; Ripabelli, G; Vitullo, M; Dallman, TJ; Pontello, M; Amar, CFL; Sammarco, ML. Gene expression in *Listeria monocytogenes* exposed to sublethal concentration of benzalkonium chloride. *Comp. Immunol. Microbiol. Infect. Dis.*, 2015, 40, 31–39.

Tenover, FC; Arbeit, RD; Goering, RV; Mickelsen, PA; Murray, BE; Persing, DH; Swaminathan, B. Interpreting chromosomal DNA restriction patterns produced by Pulsed-Field Gel Electrophoresis: criteria for bacterial strain typing. *J Clin. Microbiol.*, 1995, 33, 2233-2239.

Tien, YY; Wang, YW; Tung, SK; Liang, SY; Chiou, CS. Comparison of multilocus variable-number tandem repeat analysis and pulsed-field gel electrophoresis in molecular subtyping of *Salmonella enterica* serovars Paratyphi A. *Diagn. Micr. Infect. Dis.*, 2011, 69, 1-6.

Titze-de-Almeida, R; Willems, RJ; Top, J; Rodrigues, IP; Ferreira, RF; Boelens, H; Brandileone, MC; Zanella, RC; Felipe, MS; van Belkum, A. Multilocus variable-number tandem-repeat polymorphism among Brazilian *Enterococcus faecalis* strains. *J. Clin. Microbiol.*, 2004, 42, 4879-4881.

Torpdahl, M; Skov, MN; Sandvang, D; Baggesen, DL. Genotypic characterization of *Salmonella* by multilocus sequence typing, pulsed-field gel electrophoresis and amplified fragment length polymorphism. *J. Microbiol. Meth.*, 2005, 63, 173-184.

Tyler, KD; Wang, G; Tyler, SD; Johnson, WM. Factors affecting reliability and reproducibility of amplification-based DNA fingerprinting of representative bacterial pathogens. *J. Clin. Microbiol.*, 1997, 35 (2), 339–346.

U.S. Food and Drug Administration (FDA). Outbreaks associated with Fresh Produce: Incidence, growth, and survival of pathogens in fresh and fresh-cut produce. 2013. http://www.fda.gov/Food/FoodScienceResearch/SafePracticesforFoodProcesses/ucm091268.htm Last updated: 24/04/2013

van Belkum, A; Scherer, S; van Alphen, L; Verbrugh, H. Short-sequence DNA repeats in prokaryotic genomes. *Microbiol. Mol. Biol. Rev.*, 1998, 43, 275-293.

van Belkum, A; Tassios, PT; Dijkshoorn, L; Haeggman, S; Cookson, B; Fry, NK; Fussing, V; Green, J; Feil, E; Gerner-Smidt, P; Brisse, S; Struelens, M. Guidelines for the validation and application of typing methods for use in bacterial epidemiology. *Clin. Microbiol. Infect. Dis.*, 2007, 13, 1-46.

van der Oost, J; Jore, MM; Westra, ER; Lundgren, M; Brouns, SJ. CRISPR-based adaptive and heritable immunity in prokaryotes. *Trends Biochem. Sci.*, 2009, 34, 401-407.

van der Veen S, Abee T. Importance of SigB for *Listeria monocytogenes* static and continuous-flow biofilm formation and disinfectant resistance. *Appl. Environ. Microbiol.*, 2010b, 76(23), 7854–60.

van der Veen, S; Hain, T; Wouters, JA; Hossain, H; de Vos, WM; Abee, T; Chakraborty, T; Wells-Bennik, MHJ. The heat-shock response of *Listeria monocytogenes* comprises genes involved in heat shock, cell division, cell wall synthesis, and the SOS response. *Microbiology,* 2007, 153, 3593–3607.

van der Veen, S; van Schalkwijk, S; Molenaar, D; de Vos, WM; Abee, T; Wells-Bennik, MHJ. The SOS response of *Listeria monocytogenes* is involved in stress resistance and mutagenesis. *Microbiology*, 2010a, 156, 374–384.

van Loo, IHM; Heuvelman, KJ; King, AJ; Mooi, FR. Multilocus sequence typing of *Bordetella pertussis* based on surface protein genes. *J. Clin. Microbiol.,* 2002, 40, 1994-2001.

van Stelten, A; Simpson, JM; Ward, TJ; Nightingale, KK. Revelation by single nucleotide polymorphism genotyping that mutations leading to a premature stop codon in inlA are common among *Listeria monocytogenes* isolates from ready-to-eat foods but not human listeriosis cases. *Appl. Environ. Microbiol.*, 2010, 76, 2783–2790.

van Schaik, W; Abee, T. The role of sigma B in the stress response of Gram-positive bacteria – targets for food preservation and safety. *Curr. Opin. Biotechnol.*, 2005, 16:218–224.

Vogel, BF; Jorgensen, LV; Ojeniyi, B; Huss, HH; Gram, L. Diversity of *Listeria monocytogenes* isolates from cold-smoked salmon produced in different smokehouses as assessed by random amplified polymorphic DNA analyses. *Int. J. Food Microbiol.*, 2001, 65, 83–92.

Volpe Sperry, KE; Kathariou, S; Edwards, JS; Wolf, L. A. Multiple-Locus Variable-Number Tandem-Repeat Analysis as a tool for subtyping *Listeria monocytogenes* strains. *J. Clin. Microbiol.*, 2008, 46 (4), 1435–1450.

Wang, G; Qian, W; Zhang, X; Wang, H; Ye, K; Bai, Y; Zhou, G. Prevalence, genetic diversity and antimicrobial resistance of *Listeria monocytogenes* isolated from ready-to-eat meat products in Nanjing, China. *Food Control*, 2015, 50, 202-208.

Wemekamp-Kamphuis, HH; Wouters, JA; de Leeuw PPLA; Hain T; Chakraborty, T; Abee, T. Identification of Sigma factor B-controlled genes and their impact on acid stress, high hydrostatic pressure, and freeze survival in *Listeria monocytogenes* EGD-e. *Appl. Environ. Microbiol.*, 2004a, 70 (6), 3457–3466.

Wemekamp-Kamphuis, HH; Sleator, RD; Wouters, JA; Hill, C; Abee T. Molecular and physiological analysis of the role of osmolyte transporters BetL, Gbu, and OpuC in growth of *Listeria monocytogenes* at low temperatures. *Appl. Environ. Microbiol.*, 2004b, 70, 2912–2918.

Wernars, K; Boerlin, P; Audurier, A; Russell, EG; Curtis, GDW; Van der Mee-Marquet, N. The WHO multicenter study on *Listeria monocytogenes* subtyping: random amplification of polymorphic DNA (RAPD). *Int. J. Food Microbiol.*, 1996, 32,325–341.

Wonderling, LD; Wilkinson, BJ; Bayles, DO. The *htrA* (*degP*) gene of *Listeria monocytogenes* 10403S is essential for optimal growth under stress conditions. *Appl. Environ. Microbiol.*, 2004, 70, 1935.

Wu, S; de Lencastre, H; Tomasz, A. Sigma-B, a putative operon encoding alternate sigma factor of *Staphylococcus aureus* RNA polymerase: molecular cloning and DNA sequencing. *J. Bacteriol.*, 1996, 178, 6036–6042.

Yadav, MM; Roy, A; Bhanderi, B; Jani, RG. Prevalence of *Listeria* species including *L. monocytogenes* from apparently healthy animals at Baroda Zoo, Gujarat State, India. *JoTT.*, 2011, 3(7), 1929–1935.

Yang, M; Shao, S; Xiao, J; Wang, Q; Zhang, Y. Phylogenetic investigation of *Edwardsiella tarda* with multilocus sequence typing (MLST) and pulsed field gel electrophoresis (PFGE) typing methods. *Aquaculture*, 2013, 410-411, 79-85.

Yazdankhah, SP; Lindstedt, BA; Caugant, DA. Use of variable-umber tandem repeats to examine genetic diversity of *Neisseria meningitidis*. *J. Clin. Microbiol.*, 2005, 43, 1699-1705.

Yura, T; Nakahigashi, K. Regulation of the heat-shock response. *Curr. Opin. Microbiol.*, 1999. 2, 153-158.

Žgur-Bertok, D. DNA damage repair and bacterial pathogens. *PLoS Pathog.*, 2013, 9(11): e1003711.

Zhang, W; Jayarao, BM; Knabel, SJ. Multi-virulence-locus sequence typing of *Listeria monocytogenes*. *Appl. Environ. Microbiol.,* 2004, 70, 913–920.

Zunabovic, M; Domig, KJ; Pichler, I; Kneifel; W. Monitoring transmission routes of *Listeria* spp. in smoked salmon production with repetitive element sequence–based PCR techniques. *J. Food Prot.,* 2012, 75, 504–511.

In: Listeria monocytogenes
Editor: Tiffany Vicario

ISBN: 978-1-63483-765-1

Chapter 2

LISTERIA MONOCYTOGENES: GROWTH IN BIOFILMS IN THE FOOD INDUSTRY

***Pierluigi Di Ciccio*[1]*, Domenico Meloni*[2] *and Adriana Ianieri*[1]**

[1]Department of Food Science, University of Parma, Parma, Italy
[2]Department of Veterinary Medicine, University of Sassari, Sassari, Italy

ABSTRACT

Listeria monocytogenes is an important foodborne pathogen not because it causes large numbers of symptomatic cases but because of its relatively high case–fatality rate (20-30%). Listeriosis appears to be primarily a foodborne infection and is particularly a problem on foods that are not cooked, including ready-to-eat (RTE) meats, soft cheeses, and unpasteurized dairy products, as well as vegetables. Thermal processing of milk and meat products will destroy *L. monocytogenes* but post-processing contamination does occur. *L. monocytogenes* that contaminates foods may originate in soils from farms, in slaughtering facilities, in food processing plants and other equipments used in food manufacture. Elimination of Listeria from many environmental sources can be challenging because of the resistant, persistent biofilms formed by this microorganism. Bacterial biofilms are generally described as surface-associated bacterial communities forming microcolonies surrounded by a matrix of exopolymers. Biofilm formation can be separated into four sequential steps: (*i*) conditioning of the surface, (*ii*) initial adherence, (*iii*) physical irreversible adherence that involves the production of exopolymers that fix the cells, and (*iv*) growth of the microorganisms, which form microcolonies, and coaggregation leading to establishment of the biofilm. Such structures are dynamic systems in which cells grow, die, and/or are released, allowing contamination of other surfaces. Moreover, biofilms confer to *L. monocytogenes* a greater resistance to antimicrobial agents and they allow microorganisms to resist to environmental changes and cleaning/disinfection treatments. In the following, through an up-to-date review of (personal and non) published data, the growth of *L. monocytogenes* in biofilm in the food industry will be discussed.

1. Morphological, Physiological and Epidemiological Features of *Listeria monocytogenes*

The genus Listeria includes fifteen species: *L. monocytogenes, L. innocua, L. seeligeri, L. welshimeri, L. ivanovii, L. grayi, L. marthii, L. rocourtiae, L. leichmannii, L. weihenstephanensis, L. floridensis, L. aquatic, L. cornellensis, L. riparia* and *L. grandensis* (Meloni, 2014). Only *L. monocytogenes* and *L. ivanovii* are pathogenic, but *L. ivanovii* is strictly described as an animal pathogen (Mead et al. 1999; Buchrieser, 2007). *L. monocytogenes* is a non-sporeforming, and small Gram-positive rod of 0.5–4 µm in diameter and 0.5–2 µm in length (Meloni, 2014). *L. monocytogenes* is motile by means of peritrichous flagella at room temperature (20–25°C) and exhibits a characteristic tumbling motility. Based on somatic (O) and flagellar (H) antigens, *L. monocytogenes* has been differentiated into 13 serotypes: 1/2a, 1/2b, 1/2c, 3a, 3b, 3c, 4a, 4ab, 4b, 4c, 4d, 4e and 7 (Liu, 2006; Jamali, Radmehr and Thong, 2013). *L. monocytogenes* is catalase-positive, oxidase-negative and can grow both under aerobic and anaerobic conditions over a temperature range from 0.4°C to 45°C (Junttila, Niemela and Hirn, 1988; Walker, Archer and Banks, 1990). The optimum growth temperature is around 30–37°C. *L. monocytogenes* can grow over a pH range from 4.0 to 9.6 (optimum pH between 6 and 8) (Farber and Peterkin, 1991) and at water activity (aw) values of even 0.90 (Farber, Coates and Daley, 1992). *L. monocytogenes* is the etiologic agent of listeriosis and the majority of the infections are thought to be food-borne. Two forms of listeriosis have been described in humans and symptoms vary from a febrile gastroenteritis syndrome in healthy people (Piana et al. 2005) to sever invasive listeriosis including abortion, sepsis, and meningoencephalitis in groups at highest risk, as young, old, pregnant and immune-compromised (YOPI) people (De Cesare et al. 2007; Mackiw et al. 2016). Listeriosis is an important disease in Europe: it is the fourth most common zoonotic disease: in 2013 the EU notification rate was 0.44 cases per 100,000 and 1763 confirmed human cases were reported, mostly domestically acquired (EFSA, 2015). An 7.3% increase was observed compared with 2012 (EFSA, 2015). Listeriosis has the highest hospitalization rate cases of all zoonoses under EU surveillance: 99.1% of the cases with supplementary long-term sequelae and an estimated case fatality rate of 15.6% (EFSA, 2015).

2. Biofilm Formation by *Listeria monocytogenes*

More than 60 years after the first report on biofilms (Zobell, 1943), they are still a concern in a broad range of areas and specifically in the food, environmental and biomedical fields (Simoes et al. 2010). The term biofilm was created to describe the sessile form of microbial life, characterized by adhesion of microorganisms to biotic or abiotic surfaces, with consequent production of extracellular polymeric substances (Nikolaev and Plakunov, 2007). Anyway, the definition of biofilm has evolved significantly since its discovery and researchers are still debating a common definition. Currently, the definition of Donlan and Costerton (2002) remains the most appreciated. These authors defined the biofilm as a structured community of microbial cells, enclosed in a self produced polymeric matrix, and adherent to a surface, to interface, and to each other. However, it is now generally accepted that bacteria grow preferentially as biofilms, that are complex communities growing on a surface and surrounded by extracellular substances (Sutherland, 2001; Hall-Stoodley and Stoodley, 2005). In nature and food systems, microorganisms get attracted to solid surfaces and, initially, they are simply deposited. Later they get attached, grow and actively multiply to form a colony of cells (Allison and Sutherland, 1987). These mass of cells further become large enough to entrap organic and inorganic debris and other microorganisms leading to the formation of a microbial biofilm. The film may be a few micrometres or several millimetres thick and contains 90–97% water (Sutherland et al. 2004). Biofilms may be composed of a single species or represent a consortium of numerous species. Several microorganisms, both spoilage and pathogenic, in fact, are capable of participating in the adhesion processes and biofilm formation. In details, biofilm formation is dependent on a wide range and combination of surface morphologies such as chemistry, topography, physicochemistry, environmental conditions (pH, nutrient availability, temperature, fluid dynamics) and microbiological factors (gram negative/positive, microbial shape, structure, molecular composition, species, physico-chemistry, growth phase, age, presence of flagella, pili, capsules or exopolymeric substances) (Campoccia et al. 2013). In the case of *L. monocytogenes*, the electrostatic charge of bacterial cell walls (conferred by peptidoglycan anionic teichoic acids) and cell surface hydrophobicity (enhanced by the presence of lactic acid) have been shown to govern its attachment to stainless steel (Briandet et al. 1999a; 1999b). However, before a cell can bind to a surface, the surface is conditioned by adsorbing molecules from the surrounding environment. In brief, bacteria along with other organic and inorganic molecules gets adsorbed to the surface and, then, the accumulation of molecules at the solid-liquid interface on food-contact surfaces leads to a higher concentration of nutrients compared to the fluid phase and acts as a conditioning film (Hood and Zottola, 1997). This conditioning film favours biofilm formation and alters the physico-chemical properties of the surface, *e.g.,* changes in hydrophobicity and electrostatic charges (Dickson and Koohmaraie, 1989). The development of biofilms can occur on almost any surface in any environment in which viable microorganisms are present. In details, the biofilm formation is a stepwise and dynamical process consisting of (*i*) initial attachment, (*ii*) irreversible attachment, (*iii*) microcolony formation, (*iv*) maturation, and (*v*) dispersion. The bacteria's initial attachment is via weak electrostatic and van der Waals' forces and occurs rapidly, but, within a short time the growth of the film begins, with physical attachment of the cells to the surface by complex polysaccharide (Zottola, 1994; Palmer et al. 2007). During this stage, bacteria can easily be removed. The irreversible attachment of cells happens as the following step. A direct contact between cells and surface is facilitated by the production of surface appendages by the bacteria (flagella, fimbriae, pili) and exopolysaccharide fibrils (Jones and Isaacson, 1983). In this process, the removal of cells requires much stronger forces. The irreversibly attached bacteria grow and divide, leading to the formation of microcolonies of bacteria. During this period, the attached cells also produce additional EPS which helps in the anchorage of the cells to the surface (Costerton et al. 1995). The biofilm maturation is the step where it develops into an organized structure which can be flat or mushroom-shaped, depending on the nutrient source it is reliant upon (Chmielewski & Frank, 2003). Bacteria grow under sessile form in heterogeneous complex-enclosed microcolonies scattered with open water channels (Costerton et al. 1994; Davey and O'Toole, 2000). Dispersion is the last step in the biofilm formation cycle, and it allows the cells to revert into their planktonic form. After the formation of the biofilm, the attached bacteria are released. Detachment seems to be an active process which allows for the colonization of new niches (Flint et al. 1997; Sauer et al. 2002).

From a bacterial cell point of view, biofilm formation provides numerous advantages. Bacterial biofilm formation could be driven by at least four interconnected teleonomic values: (*i*) protection from stressful/harmful environmental conditions, by providing a certain degree of shelter and homeostasis, (*ii*) competition for and appropriation of available nutrients in a delimited area, (*iii*) benefits of metabolic interactions between microbial species from commensalism, cooperation to mutualism, and (*iv*) gene transfer enabling acquisition of new adaptative phenotypic traits (Davey and O' Toole, 2000; Molin and Tolker-Nielsen, 2003). Moreover, biofilms form a reservoir of contamination that persists where cleaning of the manufacturing plant is ineffective (Sharma and Anand, 2002). Studies have shown the capacity of several food-borne pathogens, such as *L. monocytogenes,* to persist in the environment for years Lunden et al. 2002). In particular, environmental surfaces such as floors and walls may be indirect sources of contamination *e.g.,* transference to the food products by vectors such as air, people and cleaning systems (Holah, 1992; Gibson et al. 1999). In food systems, improperly cleaned and sanitized equipment and air-borne microflora are usually considered to be the major sources of contamination (Simoes et al. 2010). Other common sources involved in biofilm accumulation are the floors, waste water pipes, bends in pipes, rubber seals, conveyor belts, stainless steel surfaces. Buna-N and Teflon seals have also been implicated as important sites for biofilm formation (Blackman and Frank, 1996; Rodriguez et al. 2007). Herald and Zottola (1988) observed the attachment of *L. monocytogenes* to stainless steel and produced attachment fibrils. The pathogen also attached to glass, polypropylene and rubber (Mafu et al. 1990) and produced a sanitizer-resistant biofilm on glass, stainless steel and Buna-N rubber surfaces (Frank and Koffi, 1990; Lee and Frank, 1991; Rönner and Wong, 1993; Lopez et al. 2008). According to Chae et al. (2006), the occurrence of foodborne outbreaks as well as sporadic cases caused by this bacterium, can be attributed to its increased ability of surviving in food processing environments through biofilm formation. Some researchers (Sasahara and Zottola 1993; Jeong and Frank 1994a and 1994b) found that *L. monocytogenes* grow preferably as a biofilm as part of a consortium of bacterial species. Kalmokoff et al. (2001) reported that a majority of *L. monocytogenes* strains might not form biofilms in monoculture. Other reports examining longer-term biofilm formation have noted that *L. monocytogenes* is a poor organism for cell attachment and biofilm formation and this has led to the suggestions that these strains may use a primary colonizing bacterium of a different species to form a biofilm consortium on a surface (Sasahara and Zottola, 1993; Carpentier and Chassaing 2004; Moltz and Martin, 2005; Gandhi and Chikindas, 2007). Several researchers reported that *L. monocytogenes* is not capable of forming thick biofilms made up of several layers (9 to 12 Log CFU/cm^2), but rather of adhering to surfaces at levels ranging from 4 to 6 Log CFU/cm^2 (Gram et al. 2007, Oliveira et al. 2010). The ability of *L. monocytogenes* to produce biofilm significantly differs according to growth temperatures, and growth surface. The ability to form biofilms at 4, 12, and 22°C results significantly more pronounced on glass than on polystyrene and stainless steel. Further, at 37°C both stainless steel and glass allows comparable amounts of biofilm, significantly higher than polystyrene (Di Bonaventura et al. 2008). The ability of *L. monocytogenes* to colonize surface at low temperatures used in the food industry, added to the recent finding that *L. monocytogenes* cells in biofilms could survive storage at 4° or 10°C for at least 5 days (Somers and Wong, 2004). Recent studies reported significant differences in biofilm formation, on PVC (Borucki et al. 2003; Djordjevic et al. 2002) or stainless steel (Folsom et al. 2006), between different *L. monocytogenes* lineages, although findings were opposite (Djordjevic et al. 2002; Chae et al. 2006). Di Bonaventura et al. (2008) did not find any relationship between phylogeny and the ability to form biofilm on different surfaces at different temperatures. Further, no differences among environmental and food strains have been observed in the experiment. At the same time, differences in biofilm production has been shown (Kalmokoff et al. 2001): serotype 1/2c forms higher average amount of biofilm than serotype 1/2a on stainless steel, and serotypes 1/2a and 4b on glass, at 37°C. Food contact surfaces have different physicochemical characteristics: hydrophilic (glass, stainless steel) and hydrophobic (polystyrene). Some studies have failed to find a correlation between hydrophobicity and attachment to surface. Particularly, Chae et al. (2006) recently found that attachment on glass is independent of hydrophobicity level. Other authors (Di Bonaventura et al. 2008) suggested that hydrophobicity level is correlated with biofilm formation on glass. These conflicting results are probably due to the fact that hydrophobicity can differ between serotypes or strains, and can change with variation in growth conditions (Briandet et al. 1999; Giovannacci et al. 2000). The cell surface is generally considered a significant factor in bacterial attachment to surfaces. Many studies suggest that microbial cell surface charge and hydrophobicity play an important role in the initial steps of microbial adhesion. The hydrophobicity and surface charges of bacteria can differ between species, serotypes or strains, and can change with variation in growth conditions, physiological state of cells, and composition of suspension media (Briandet et al. 1999; Giovannacci et al. 2000). Chae et al. (2006) demonstrated that high levels of extracellular carbohydrates produced by *L. monocytogenes* increased their ability to form biofilms, indicating the importance of this characteristic for a given strain's biofilm forming ability. Bacterial attachment to surfaces is influenced not only by cell surface and hydrophobicity but also by the presence of particular surface appendages such as flagella and fimbriae. It has been suggested that flagellum-mediated motility is required for initial cell attachment during biofilm formation by overcoming any repulsive interfacial forces (Herald and Zottola, 1988; O'Toole and Kolter. 1998; Kalmokoff et al. 2006). Flagella, in addition to being the locomotive organelles of bacteria, have also reported to serve as an adhesive structure (Moens and Vanderleyden, 1996). At this regard, Vatanyoopaisarn et al. (2000) found that flagella per se, not motility, facilitate the early stages of attachment of *L. monocytogenes* to stainless steel. Very little is known about relationship between motility and biofilm formation and, in any case, only swimming motility has been considered. Recent findings suggest that Gram-positive organisms, including *L. monocytogenes*, have evolved multiple molecular strategies for the formation of pili on microbial surfaces (Hung and Schneewind, 2004). Further, Dons et al. (2004) showed that motility by swarming is critical in host-cell invasion and virulence of *L. monocytogenes*. In an attempt to individuate a relationship between *L. monocytogenes* flagellum-mediated motility (swimming) and biofilm formation, it has been shown that the temperature strongly influences flagellum production in *Listeria* spp. (Di Bonaventura et al. 2008). On the contrary, motility did not show a positive correlation with biofilm-forming ability, regardless of the substratum (Djordjevic et al. 2002). On the whole, functionally active flagella (motility) are probably not required for biofilm formation in *L. monocytogenes* and changes in surface structures other than flagella contribute to attachment of the bacterium to solid surface. At both 4° and 12°C *L. monocytogenes* produces, regardless of substratum, a rudimentary biofilm consisting only of sparse clusters of cells and minimum amounts of EPS. On the contrary, at 22° and 37°C biofilm is more complex, in terms of cell number and EPS produced, although on glass only it appeared as a complex three-dimensional architecture, consisting of dense aggregates of cells held together by extracellular matrix and surrounded by void areas, probably representing water channels for nutrient circulation and facilitation of waste removal in biofilm (Costerton et al. 1995; Donlan and Costerton, 2002). The relevant amount of EPS that can be observed in biofilms formed on glass raises a relevant hygienic/sanitary issue, since the chlorine resistance of *L. monocytogenes* strains may be closely associated with the amount of EPS produced by the biofilm cell (Folsom and Frank, 2006). The time available for biofilm formation will depend on the frequency of cleaning regimes. Product contact surfaces may typically be cleaned several times per day, while environmental surfaces such as walls may be cleaned once per week. There is, therefore, more time for biofilm formation on environmental surfaces. Gibson et al. (1999) found that although attachment to a variety of surfaces in the food processing environment readily occurred, extensive surface colonization and biofilm formation only occurred on environmental surfaces. Finally, routine clean in place (CIP) cycles in food plants may not remove all cells and the remaining biofilm cells may allow more rapid recolonisation of the plant by a process of seeding or by providing a surface to which new cells may readily adhere (Mashall, 1994). The situation is exacerbated if fouling has occurred and is not removed during CIP (Hinton et al. 2002).

3. Biofilm Formation in Dairy Factories

Raw milk contaminated with *L. monocytogenes* could represent a potential source for the introduction of the pathogen in dairy products processing plant environment (Vitas et al. 2004; AlTahiri et al. 2008; Ibba et al. 2013). The frequent presence of *L. monocytogenes* in bulk tank milk and in in-line milk filter samples (Latorre et al. 2009; 2010) and the nature of raw milk compounds contribute to the establishment of biofilms by *L. monocytogenes* strains on the surfaces of milking equipment in the dairy farms (Latorre et al. 2010). As milk is a complex substrate for microorganisms, the formation of biofilms on various surfaces such as glass, PVC or stainless steel (Norwood and Gilmour, 2001; Djordjevic et al. 2002; Chae et al. 2006) is rapid for *L. monocytogenes* strains which can easily multiply in milk residues after improper leaning of milk equipment (Hayes et al. 2001). Accumulation of organic material debris in the milking machines may create appropriate conditions for bacterial growth (Murphy and Boor, 2000) and could also help in the attachment of bacteria by creating a conditioning film (Zottola and Sasahara, 1994). The presence of *L. monocytogenes* in raw milk could also pose a potential risk of contamination for a milk processing plant (Waak et al. 2002) and highlights the need to prevent the biofilm establishment in milking equipment (Latorre et al. 2010). Hygiene in the milking routine, correct implementation of milking equipment cleaning protocols and replacing plastic and other materials in milking equipment would help to prevent the presence of *L. monocytogenes* in milk processing plants and the contamination of dairy products (Dalton et al. 1997; Koch et al. 2010; Latorre et al. 2010). The presence of *L. monocytogenes* has been well documented in the dairy industry and the ability to form biofilms (Harvey et al. 2007) may contribute to its persistence in milk processing plants. Persistence in processing plant environments is generally associated with the presence of resident strains that colonise specific niches of improper hygienic design (e.g., hard-to-clean recesses, crevices, worn material) and/or improperly cleaned (Blackman and

Frank, 1996; Latorre et al. 2010; Ibba et al. 2013). Niches are represented by areas with favourable conditions such as moisture and condenses, low temperatures and high salt concentrations (Tompkin, 2002; Ibba et al. 2013). *L. monocytogenes* harbouring niches support post-process contamination of dairy products made from pasteurised or thermised milk (Unnerstad et al. 1996; Tompkin, 2002; Ibba et al. 2013). *L. monocytogenes* has the potential to adhere and form biofilms on materials frequently found in milk handling equipment, milk lines, or milk tanks such as stainless steel (Norwood and Gilmour, 1999; Beresford et al. 2001), rubber, or plastic (Beresford et al. 2001). This ability allows the microorganism to adapt and resist against detergents and sanitisers (Chavant et al. 2004; Pan et al. 2006). The regular implementation of environmental sampling programs is strongly recommended in order to monitor the level of *L. monocytogenes* contamination in dairy factories and the presence of persistent harborage sites (Ibba et al. 2013).

4. Biofilm Formation in Meat Processing Environments

In meat-processing environments, all surfaces and materials are likely to be colonized by *L. monocytogenes* (*e.g.,* pipelines, knifes, hooks, gaskets, conveyor belts) if sanitation procedures are inadequate and/or insufficient frequent. In addition, more the surfaces present potential harborage niches (hollow parts, crevices, cracks, unpolished or worn materials), more they are difficult to clean and disinfect and more the bacteria can colonize and occupy them for finally adapt and tolerate sanitation procedures. It is now established that when the contamination of meat products occurs, the biofilms are the major source of contamination. The meat processing environment provides a suitable environment for the *L. monocytogenes* biofilm formation. Organic residues, in fact, could be a niche for microorganism accumulation and biofilm formation as it is a source of cross-contamination (Pérez-Conesa et al. 2006; Brooks and Flint, 2008; Simões, and Vieira, 2010).

Additionally, It has been seen that at low temperatures, *L. monocytogenes* can adhere and survive on meat contact surfaces, and even increase in population as time progressed (Ryu and Beuchat, 2005; Dourou et al. 2011). *L monocytogenes* can adhere rapidly and firmly to inert surfaces commonly found in the meat processing industry (Autio et al. 1999; Lyytikainen et al. 2000; Miettinen et al. 2001; D'Orio et a., 2007). *L. monocytogenes*, in fact, is capable of adhering and forming biofilm on common materials used in meat industry such as metal, glass, rubber and plastic surfaces (Hood and Zottola, 1997; Chae and Schraft, 2000, 2006; Sinde and Cerballo, 2000; Joseph et al. 2001; Stepanovic et al. 2004; Moltz and Martin, 2005).

The mechanisms by which *L. monocytogenes* survive under harsh conditions of physical and chemical stress are, at least in part, due to the its ability to form biofilms on surfaces within the food processing environment (Wong, 1998; Holah et al. 2004; Chae et al. 2006). Usually, Listeria sources in meat processing plants are conveyor belts, cutters, slicers, bringing and packaging machines, coolers and freezers as well as floors and drains (Harvey et al. 2007).

5. Biofilm Formation in Seafood Factories

L. monocytogenes, as a psychotrophic bacterium, is widespread in food processing facilities and has been isolated from different processing environments, including seafood factories (Dauphin et al. 2001; Norton et al. 2001; Vogel et al. 2001a; Hoffman et al. 2003). When introduced into seafood processing environments, *L. monocytogenes* can tolerate and continue to grow in conditions (e.g., refrigeration temperatures and high salt levels) that prevent the growth of many other foodborne pathogens. *L. monocytogenes* survives extremely well in the seafood processing plant environment: can adapt to disinfectants and adhere to various surfaces. "In-house" *L. monocytogenes* flora can contaminate finished products during processing (Rorvik et al. 2000; Vogel et al. 2001b; Autio et al. 2003; Miettinen & Wirtanen, 2005; Arevalos-Sánchez et al. 2012). *L. monocytogenes* may be introduced into seafood processing plants through a variety of routes, including raw materials, employees' shoes or clothes, and equipment (boxes, crates, carts). Once introduced into the processing plants, it is able to survive and persist for a long time under adverse conditions (Farber and Peterkin, 1991; Gram et al. 2007; Gandhi and Chikindas, 2007) forming biofilms when resident populations become established in niches in the plant (D'Orio et al. 2007; Di Bonaventura et al. 2008). Several authors have reported the colonisation of some subtypes of *L. monocytogenes* in seafood processing plants, equipment, utensils and brine causing persistent contamination for months or even years (Rørvik et al. 1995; López et al. 2008). Apart from the processing environment, the raw materials have been regarded as a source of contamination (Gudmundsdottir et al. 2005). There are indications that *L. monocytogenes* present in the raw material can proliferate and contaminate the final products, especially those that are not heat-treated before consumption (Vogel et al. 2001b; Eklund et al. 2004; Miettinen and Wirtanen, 2005). These resident populations and the biofilms they form to enhance their survival are not easily eliminated by general-purpose cleaners or sanitizers and normal sanitation procedures (Di Ciccio et al. 2012). Fingerprinting techniques have significantly contributed to an improved understanding of the ecology, sources, and spread of *L. monocytogenes* in seafood processing plant environments (Johansson et al. 1999; Autio et al. 1999; Markkula et al. 2005; Di Ciccio et al. 2012). While a diversity of different *L. monocytogenes* strains are found in most seafood processing plants, individual processing facilities often harbour unique *L. monocytogenes* populations and strains, which persist for months or years in the plant or its product despite sanitation protocols designed to eliminate them (Autio et al. 1999; Dauphin et al. 2001; Norton et al. 2001; Vogel et al. 2001b; Hoffman et al. 2002; Di Ciccio et al. 2012). Patterns of persistent processing plant contamination have been reported for a variety of seafood processing environments (Ben Embarek, 1994; Rørvik et al. 1995; Norton et al. 2001; Medrala et al. 2003; Thimothe et al. 2004; Vitas and Garcia-Jalon, 2004; Garrido et al. 2008; Cruz and Fletcher, 2011). These findings indicate that, while a variety of *L. monocytogenes* may be introduced (probably daily) into the plant environment from different sources, most are eliminated by cleaning and sanitation and replaced by "in-house" flora. (Di Ciccio et al. 2012). Some subtypes appear to colonize specific niches in the plant environment and persist over time. Thus monitoring for the presence and reintroduction of persistent *L. monocytogenes* contamination should be a component of every control strategy (Tompkin et al. 1999, 2002; Hoffman et al. 2003; Di Ciccio et al. 2012). Persistent *L. monocytogenes* contamination in seafood processing plants represents a major concern for the industry and public health. Some studies using molecular subtyping of *L. monocytogenes* isolates specifically showed that the subtypes persisting in respective plants were responsible for the majority of finished product contamination (Farber and Peterkin, 1991; Jemmi and Keusch, 1992; Rorvik et al. 1995; Nilsson, Huss and Gram 1997; Norton et al. 2001; Corcoran et al. 2006; Meloni et al. 2009; Miya et al. 2010 ; Di Ciccio et al. 2012; Rotariu et al. 2014). Eradication of persistent strains in the plant will reduce the risk of finished product contamination from environmental sources (Autio et al.1999).

6. Emerging Strategies to Control Biofilm in the Food Industry

The first and most important thing to do is to pevent biofilm formation by regularly cleaning and disinfecting so as to not allow the cells to firmly attach to contact surfaces (Midelet and Carpentier, 2004). Therefore, research for new strategies in biofilm control is an important area of focus. The growing negative consumer perception against artificial synthetic chemicals, however, has shifted this research effort toward the development of alternatives, environmental friendly disinfectants, which consumers perceive as "naturals." The use of enzyme-based detergents as bio-cleaners can serve as a viable option to overcome the biofilm problem in the food industry. Due to the EPS heterogeneity, a mixture of enzymes may be necessary for sufficient biofilm degradation. The use of enzymes in biofilm control is still limited due to the low prices of the chemicals used today compared with the costs of the enzymes. In fact, the technology and production of these enzymes and the enzyme-based detergents are mostly patent-protected. Moreover, the low commercial accessibility of different enzyme activities limits their current usage (Johansen et al. 1997). Enzymes and detergents have also been used as synergists to improve disinfectant efficacy (Johansen et al. 1997; Parkar et al. 2004). Another innovative approach to sensitizing industrial biofilms to biocides is to use hyper-swimming tunnelling bacteria. A recent study demonstrated that planktonic bacilli propelled by flagella were able to tunnel deep into a biofilm structure (Houry et al. 2012). The transient pores created in the matrix by these bacterial stealth swimmers increased macromolecular transfer within the biofilm and thus enhanced the killing of biofilm cells by facilitating the penetration and action of disinfectants from the environment (Houry et al. 2012). An interesting study on the control of *L. monocytogenes* in a biofilm by competitive exclusive microorganisms indicated that *L. monocytogenes* in biofilms can be strongly inhibited ($> 5 \log^{10}$ CFU of *L. monocytogenes*/cm^2) by metabolites of *Enterococcus durans* and *Lactococcus lactis subsp. lactis*. These two strains were isolated and screened from biofilms obtained from the floor drains of food processing plants that were free of *L. monocytogenes* in records. *Enterococcus durans* and *Lactococcus lactis, subsp. lactis* are beneficial organisms and can be used as starter cultures for food fermentation and they have a great potential for the control of *L. monocytogenes* in biofilms in food processing environment (Zhao et al. 2004). Studies have shown that some detergents are bactericidal and some disinfectants may even depolymerize EPS, thus enabling the detachment of biofilms from surfaces, *e.g.,* oxidants such as chlorine and hydrogen peroxide (Juven and Pierson, 1996). Monolaurin (glycerol monolaurate) was also found to be lethal to *L. monocytogenes* at low concentrations. In addition, a synergistic interaction between monolaurin and organic acids like acetic acid also caused a pronounced inhibition of *L. monocytogenes* (Oh and Marshall, 1996). More recently, several authors (García-Almendárez et al. 2008; Guerrieri et al. 2009; Speranza et al. 2009) demonstrated that bacteriocin producers showed the best antilisterial potentiality. Microbial molecules, commonly used as biopreservatives, such as nisin, lauricidin, reuterin and pediocin, have been well documented for their biofilm control potential against microorganisms commonly found in dairy processing facilities, including *L. monocytogenes* (Dufour et al. 2004; Zhao et al. 2004; Mahdavi et al. 2007; Garcia-Almendarez et al. 2008). Application of bacteriocins have also been tried out on food packaging materials for the biocontrol of *L. monocytogenes* on meats (Ming et al. 1997). The antagonistic effect was made more effective by the decrease of pH due to lactic acid production in the *L. plantarum* biofilm, outcome confirmed by the considerable activity in *L. plantarum* biofilm and suspension. Comparing the antilisterial activity of LAB biofilm against both planktonic and adherent cells, the *L. monocytogenes* adherent cells showed a higher resistance. This finding can be explained by the greater resistances that microorganisms display to exogenous agents in biofilms (Sandasi et al. 2008). The use of bacteriophages in order to control the biofilms may provide a natural, highly specific, non-toxic, feasible approach for controlling several microorganisms involved in biofilm formation (Kudva et al. 1999). A bacteriophage (*L. monocytogenes* phage ATCC 23074-B1) was used successfully in *L. monocytogenes* biofilm inactivation (Hibma., 1997; Lu and Collins, 2007). This enzymatic phage had the ability to attack the bacterial cells in the biofilm and the biofilm matrix, substantially reducing the biofilm cell counts (more than 99.9% of removal). The technology for this has not yet been successfully developed and relatively little information is available on the action of bacteriophages on biofilms (Sillankorva et al. 2004; Sutherland et al. 2004). Production of siderophores is a virulence factor in many microorganisms, acting as biocontrol molecules (Gram et al. 1999). A pioneer study indicated that siderophore-containing *Pseudomonas spp.* culture supernatants inhibited growth of *Shewanella putrefaciens*, as did the addition of iron-chelators (Gram, 1993). Such biological mechanisms, alone or as part of synergistic procedures could provide a new line of efficient biofilm control strategies (Banin et al. 2005). In the particular case of *L. monocytogenes*, iron availability affects several bacterial properties. An iron-deficient growth leads to a decrease in this bacterium surface hydrophobicity, together with major alterations in the surface protein composition (Conte et al. 1996). Recently, the use of chlorine dioxide (CD) as decontamination agent has demonstrated its high effectiveness against a wide variety of microorganisms. It is a strong oxidizing agent with several advantages, such as the formation of low toxic disinfection by-products, effectiveness at low concentrations, low reaction time, ease of generation and effectiveness over a broad range of pH (Chang et al. 2000). It is conceivable that quorum sensing inhibition may represent a natural, widespread, antimicrobial strategy with significant impact on biofilm formation. A good understanding of the cell–cell signalling phenomenon of bacteria such as *L. monocytogenes* can be used to control the biofilm formation process by the identification of products that can act as quorum sensing antagonists (Simões et al. 2009). This property can lead to the development of new and efficient natural products for biofilm control. Modifications to inert surfaces so as to reduce their interactions with microbial cells and subsequent adhesion have been proposed (Simoes et al. 2010). In this regard, several attempts have been made to avoid biofilm formation by the incorporation of antimicrobial products into surface materials (Park et al. 2004), by coating surfaces with antimicrobials (Gottenbos et al. 2001) and or by modifying the surfaces physicochemical properties (Whitehead et al. 2005; Rosmaninho et al. 2007). Gottenbos et al. (2001) demonstrated a reduction in infection rate using silicone rubber implants with covalently coupled quaternary ammonium coatings. Other authors reported biofilm formation inhibition by coating surfaces with silver (Hashimoto, 2001). Cloete and Jacobs (2001) reported that surface pre-conditioning with surfactants has potential to prevent bacterial adhesion. The surfactants gave more than 90% inhibition of adhesion. More recently, other studies (Meylheuc et al. 2006; Splendiani et al. 2006) reinforced the efficiency of surfactants and surface pre-conditioning on biofilm formation control. Modified highly hydrophobic polystyrene demonstrated their antibacterial efficacy against different micro-organisms, including *L. monocytogenes* (Poncin- Epaillard et al. 2013). Surface modifications using nano-materials such as silver, cobalt and iron mixed oxides are interesting candidates for the prevention of biofouling, notably in the context of treatment technologies for drinking water (Ashraf et al. 2014). In addition to biological and/or chemical alternatives, new physical processes (*e.g.,* pulsed-light or laser decontamination devices) are promising option for specific industrial applications. Plasma gases have proved their efficacy against biofilms (Kamgang et al. 2007; Kamgang-Youbi et al. 2009; Joaquin et al. 2009). The literature demonstrates that there is no one strategy with absolute biofilm control efficiency. Nevertheless, the importance of adequate cleaning and disinfection procedures in order to avoid *L. monocytogenes* becoming established in processing environments and thus posing a threat of product contamination is widely accepted. Furthermore, continuous environmental monitoring schemes for *L. monocytogenes* are of major importance to identify potential contamination sources and as an early warning system for food business operators, especially in food-processing plants with low *L. monocytogenes* prevalence in their food products. Nowadays, the most efficient practical means for limiting microbial growth in the food processing industry includes good production hygiene, a rational running of the process line, and effective use of cleaning and disinfectant products (Singh et al. 2002). In conclusion, preventing biofilm formation would be a more logical option than treating it. In addition, the processing equipments should be designed with high standards of hygiene in mind (Di Ciccio et al. 2012).

REFERENCES

Al-Tahiri, R., Omar, S., Rewashdeh, A. (2008). A study of the occurrence of Listeria species in raw sheep milk. *Int. J. Dairy Technol.* 61:347-1.

Allison, D. G., Sutherland I. W. (1987). The role of exopolysaccharides in adhesion of freshwater bacteria. *J. Gen. Microbiol.*, 133, 1319–1327.

Arevalos-Sánchez M., Regalado, C., Martin, S.E., Domínguez-Domínguez, J., García Almendárez B.E. (2012). Effect of neutral electrolyzed water and nisin on *Listeria monocytogenes* biofilms and on listeriolys in O activity. *Food Control* 24:116-22.

Ashraf, M.A., Ullah, S., Ahmad, I., Qureshi, A.K., Balkhair, K.S., Abdur Rehman, M. (2014). Green biocides, a promising technology: current and future applications to industry and industrial processes. *J. Sci. Food Agric.* 94, 388-403.

Autio, T., Hielm, S., Miettinen, M., Sjoberg, A.M., Aarnisalo, K., Bjorkroth, J., Mattila-Sandholm, T., Korkeala, H. (1999). Sources of *Listeria monocytogenes* contamination in a cold-smoked rainbow trout processing plant detected by pulsed field gel electrophoresis typing. *Appl. Environ. Microbiol.* 65, 150-155.

Autio, T., Keto-Timonen, R., Lunden, J., Bjorkroth, J., Korkeala, H. (2003). Characterisation of persistent and sporadic *Listeria monocytogenes* strains by pulsed-field gel electrophoresis (PFGE) and amplified fragment length polymorphism (AFLP). *Syst. Appl. Microbiol.* 26, 539-545.

Banin, E., Vasil, M. L., Greenberg, P. (2005). Iron and *Pseudomonas aeruginosa* biofilm formation. *Proceedings of the National Academy of Sciences USA,* 102, 11076–11081.

Ben Embarek, P.K. (1994). Presence, detection and growth of *Listeria monocytogenes* in seafoods: a review. *Int. J. Food Microbiol.* 23, 17e34.

Beresford, M. R., Andrew, P. W., Shama, G. (2001). *Listeria monocytogenes* adheres to many materials found in food-processing environments. *J. Appl. Microbiol.* 90:1000–1005.

Blackman, I.C., Frank, J.F. (1996). Growth of *Listeria monocytogenes* as a biofilm on various food-processing surfaces. *J. Food Prot.* 59, 827–831.

Borucki, M. K., Peppin, J. D., White, D., Loge, F., Call, D. R. (2003). Variation in biofilm formation among strains of *Listeria monocytogenes. Appl. and Environ Microbiol* 69, 7336-7342.

Briandet, R., Leriche, V., Carpentier, B., Bellon-Fontaine M. N. (1999a). Effects of the growth procedure on the surface hydrophobicity of *Listeria monocytogenes* cells and their adhesion to stainless steel. *J.Food Prot.*, 62, 994-998.

Briandet, R., Meylheuc, T., Maher, C., Bellon-Fontaine, M.N. (1999b). *Listeria monocytogenes* Scott A: cell surface charge, hydrophobicity, and electron donor and acceptor characteristics under different environmental growth conditions. *Appl. Environ. Microbiol.* 65, 5328-5333.

Brooks, J. D., Flint, S. H. (2008). Biofilms in the food industry: Problems and potential solutions. *Int. J. Food Sci. and Tech.*, 43(12), 2163–2176.

Buchrieser, C. (2007). Biodiversity of the species *Listeria monocytogenes* and the genus Listeria. *Microbes Infect.* 9(10):1147-55.

Campoccia, D., Montanaro, L, Arciola, C.R (2013). A review of the biomaterials technologies for infection-resistant surfaces. *Biomaterials* 2013, 34:8533-8554.

Carpentier, B., Chassing, D. (2004). Interactions in biofilms between *Listeria monocytogenes* and resident microorganisms from food industry premises. *Int. J. Food Microbiol.*, 97, 111–122.

Chae, M.S., Schraft, H., Hansen, L.T., Mackereth, R. (2006). Effects of physicochemical surface characteristics of *Listeria monocytogenes* strains on attachment to glass. *Int. J. Food Microbiol.* 23, 250–259.

Chang C.Y., Hsieh Y.H., Shih I.C., Hsu S.S., Wang K.H. (2000). The formation and control of disinfection by-products using chlorine dioxide, *Chemosphere* 41, 1181–1186.

Chavant, P., Gaillard-Martinie, B., Hébraud, M. (2004). Antimicrobial effects of sanitizers against planktonic and sessile *Listeria monocytogenes* cells according to the growth phase. *FEMS Microbiol. Lett.* 236:241–248.

Chmielewski, R. A. N., Frank, J. F. (2003). Biofilm formation and control in food processing facilities. *Comp. Rev. Food Sci. and Food Safety*, 2, 22–32.

Cloete T.E., Jacobs, L. (2001). Surfactants and the attachment of *Pseudomonas aeruginosa* to 3CR12 stainless steel and glass, *Water SA* 27, 21–26.

Conte, M. P., Longhi, C., Polidoro, M., Petrone, G., Buonfiglio V., Di Santo S., et al. (1996). Iron availability affects entry of *Listeria monocytogenes* into the enterocytelike cell line Caco-2. *Infection and Immunity*, 64, 3925–3929.

Corcoran, D., Clancy, D., O'Mahony, M., Grant, K., Hyland, E., Shanaghy, N., Whyte, P., McLauchlin, J., Moloney, A., Fanning, S. (2006). Comparison of *Listeria monocytogenes* strain types in Irish smoked salmon and other foods. *Int.J. Hygiene Environ. Health* 209, 527–534.

Costerton, J. W., Lewandowski, Z., deBeer, D., Caldwell, D., Korber, D., James, G., (1994). Biofilms, the customized microniche. *J. of Bacteriol.*, 176, 2137–2142.

Costerton, J. W., Lewandowski, Z., Caldwell, D., Korber, D., Scott, H. (1995). Microbial biofilms. *Ann. Rev. Microbiol.*, 49, 711-745.

Cruz, C.D., Fletcher, G.C. (2011) Prevalence and biofilm-forming ability of *Listeria monocytogenes* in New Zealand mussel (*Perna canaliculus*) processing plants. *Food Microbiol.* 2011, (7):1387-93.

Dalton, C. B., Austin, C. C., Sobel, J., Hayes, P. S., Bibb, W. F., Graves, L. M., Swaminathan, B., Proctor, M. E., Griffin, P.M.. (1997). An outbreak of gastroenteritis and fever due to *Listeria monocytogenes* in milk. N. *Engl. J. Med.* 336:100–105.

Dauphin, G., Ragimbeau, C., Malle, P. (2001). Use of PFGE typing for tracing contamination with *Listeria monocytogenes* in three cold-smoked salmon processin plants. *Int. J. Food Microbiol.* 64, 51-61.

Davey, M. E, O'Toole G. A. (2000). Microbial biofilms: from ecology to molecular genetics. *Microbiol.Mol.Biology Rev.*, 64, 847– 67.

De Cesare, A., Mioni, R.; Manfreda, G. (2007). Prevalence of *Listeria monocytogenes* in fresh and fermented sausages and ribotyping of contaminating strains. *Int. J. Food Microbiol.* 2007, 120, 124–130.

Dickson, J. S., Koohmaraie, M. (1989). Cell surface charge characteristics and their relationship to bacterial attachment to meat surfaces. *Appl. and Environ.Microbiol.* 55, 832–836.

Di Bonaventura, G., Piccolomini, R., Paludi, D., D'Orio, V., Vergara, A., Conter, M., Ianieri, A. (2008). Influence of temperature on biofilm formation by *Listeria monocytogenes* on various food-contact surfaces: relationship with motility and cell surface hydrophobicity. *J of Appl. Microbiol.* 104, 1552–1561.

Di Ciccio P, Meloni D, Festino AR, Conter M, Zanardi E, Ghidini S, Vergara A, Mazzette R, Ianieri A. (2012). Longitudinal study on the sources of *Listeria monocytogenes* contamination in cold-smoked salmon and its processing environment in Italy. *Intl. J. Food Microbiol.* 158:79–84.

Djordjevic, D., Wiedmann, M., Mc Landsbourough, L.A. (2002). Microtiter plate assay for assessment of *Listeria monocytogenes* biofilm formation. *Appl and Environ Microbiol* 68, 2950–2958

Donlan, R. M. , Costerton, J. W. (2002). Biofilms: survival mechanisms of clinically relevant microorganisms. *Clin. Microbiol. Rev.* 15(2):167-93.

Dons, L., Eriksson, E., Jin, Y., Rottenberg, M. E., Kristensson K., Larsen, C. N., Bresciani, J., Olsen, J. E. (2004). Role of flagellin and the two-component CheA/CheY system of *Listeria monocytogenes* in host cell invasion and virulence. *Infection and Immunity*, 72, 3237-3244.

D'Orio, V., Paludi, D., Conter, M., Vergara, A., Ianieri, A., 2007. Biofilm formation by environmental strains of *Listeria monocytogenes*: relationship with hydrophobicity, motility and scanning electron microscopy analysis. Food, New Options for the Industry. EFFoST/EHEGD Conference 2007 – Practical applications of research results. Lisbon 14–16 November 2007, p. 8.

Dourou, D., Beauchamp, C. S., Yoon, Y., Geornaras, I., Belk, K. E., Smith, G. C., Nychas, G. J., Dufour, M., Simmonds, R. S., Bremer, P. J. (2004). Development of a laboratory scale clean-in-place system to test the effectiveness of "natural" antimicrobials against dairy biofilms. *J. Food Prot.*, 67, 1438–1443.

EFSA and ECDC (European Food Safety Authority and European Centre for Disease Prevention and Control). (2015). The European union summary report on trends and sources of Zoonoses, Zoonotic agents and food-borne outbreaks in 2013. EFSA Journal 2015, 13(1), 162. http://dx.doi.org/10.2903/j.efsa.2015.3991, 3991.

Eklund, M.W., Poysky, F.T., Paranjpye, R.N., Lashbrook, L.C., Peterson, M.E., Pelroy, G.A. (2004 Incidence and sources of *Listeria monocytogenes* in cold-smoked fishery products and processing plants. *J. Food Prot.*, 58, 5, 502-508.

Farber, J. M., Peterkin, P. I. (1991). *Listeria monocytogenes*, a food-borne pathogen. *Microbiol. Rev.*, 55, 476-511.

Farber, J. M., Coates, F., & Daley, E. (1992). Minimum water activity requirements for the growth of *Listeria monocytogenes*. *Letters in Applied Microbiology,* 15(3), 103-105.

Flint, S. H., Bremer, P. J., Brooks, J. D. (1997). Biofilms in dairy manufacturing plant - description, current concerns and methods of control. *Biofouling*, 11, 81–97.

Folsom, J. P., Frank, J. F. (2006). Chlorine resistance of *Listeria monocytogenes* biofilms and relationship to subtype, cell density, and planktonic cell chlorine resistance. *J. Food Prot.*, 69, 1292-1296.

Frank, J. F. Koffi, R.A. (1990). Surface adherent growth of *Listeria monocytogenes* is associated with increased resistance to surfactant sanitizers and heat. *J. Food Prot.*, 53, 550–554.

Fuqua, C., Greenberg, E. P. (2002). Listening in on bacteria acyl-homoserine lactone signaling. *Nat. Rev. in Mol. Cell Biol.*, 3, 685–695.

Garcia-Almendarez, B. E., Cann, I. K. O., Martin, S. E., Guerrero-Legarreta, I. Regalado, C. (2008). Effect of Lactococcus lactis UQ2 and its bacteriocin on *Listeria monocytogenes* biofilms. *Food Control*, 19, 670–680.

Garrido, V., Torroba, L., García-Jalón, I., & Vitas, A.I. (2008). Surveillance of listeriosis in Navarre, Spain, 1995–2005 – epidemiological patterns and characterisation of clinical and food isolates. *Eurosurveillance,* 13, 19058.http://www.eurosurveillance.org

Gram, L., Bagge-Ravn, D., Ng, Y.Y., Gymoese, P., Vogel, B.F. (2007). Influence of food soiling matrix on cleaning and disinfection efficiency on surface attached *Listeria monocytogenes*. *Food Control* 18, 1165-1171.

Ghandi, M., Chikindas, M.L. (2007). Listeria: a foodborne pathogen that knows how to survive. *Int. J. Food Microbiol.* 113, 1–15.

Gibson, H. J., Taylor, H., Hall, K. E., Holah, J.T. (1999). Effectiveness of cleaning techniques used in the food industry in terms of the removal of bacterial biofilms. *J. of Appl. Microbiol.*, 87, 41–48.

Giovannacci, I., Ermel, G., Salvat, G., Vendeuvre, J. L., Bellon-Fontaine, M. N. (2000). Physicochemical surface properties of five *Listeria monocytogenes* strains from a pork-processing environment in relation to serotypes, genotypes and growth temperature. *J. of Appl. Microbiol.*, 88, 992–1000.

Gottenbos, B., van der Mei, H.C., Klatter, F., Nieuwenhuis, P., Busscher, H. J. (2001). In vitro and in vivo antimicrobial activity of covalently coupled quaternary ammonium silane coatings on silicone rubber. *Biomaterials,* 23, 1417–1423.

Gram, L. (1993). Inhibitory effect against pathogenic and spoilage bacteria of Pseudomonas strains isolated from spoiled and fresh fish. *Appl. Environ. Microbiol.,* 59, 2197–2203.

Gram, L., Melchiorsen, J., Spanggaard, B., Huber, I., Nielsen, T. F. (1999). Inhibition of *Vibrio anguillarum* by *Pseudomonas fluorescens* AH2, a possible probiotic treatment of fish. *Appl. Environ. Microbiol.*, 65, 969–973.

Gram, L., Bagge-Ravn, D., Ng, Y. Y., Gymoese, P., Vogel, B. F. (2007). Influence of food soiling matrix on cleaning and disinfection efficiency on surface attached *Listeria monocytogenes*. *Food Control*, 18, 1165-1171.

Gudmundsdottir, S., Gudbjornsdottir, B., Lauzon, H.L., Einarsson, H., Kristinsson, K.G., Kristjansson, M. (2005). Tracing *Listeria monocytogenes* isolates from cold-smoked salmon and its processing environment in Iceland using pulsed-field gel electrophoresis. *Int. J. Food Microbiol.* 101, 41-51

Guerrieri, E., de Niederhäusern, S., Messi, P., Sabia, C., Iseppi, R., Anacarso, I., Bondi, M. (2009). Use of lactic acid bacteria (LAB) biofilms for the control of *Listeria monocytogenes* in a small-scale model. *Food Control*, 20, 861-865.

Hayes, M., Ralyea, R.D., Murphy, S.C., Carey, N.R., Scarlett, J.M., Boor, K.J. (2001). Identification and characterization of elevated microbial counts in bulk tank raw milk. *J. Dairy Sci.* 84, 292–298.

Hall-Stoodley, L., Stoodley, P. (2005). Biofilm formation and dispersal and the transmission of human pathogens. *Trends in Microbiology*, 13, 7–10.

Harvey, J., Keenan, K. P., Gilmour, A. (2007). Assessing biofilm formation by *Listeria monocytogenes* strains. *Food Microbiol.*, 24, 380-392.

Hashimoto H., Evaluation of the anti-biofilm effect of a new antibacterial silver citrate/lecithin coating in an in-vitro experimental system using a modified Robbins device, *J. of the Japanese Association for Infect. Dis.* 75, 678–685.

Herald, P. J., Zottola, E. A., (1988). Attachment of *Listeria monocytogenes* to stainless steel surfaces at various temperatures and pH values. *J. Food Prot.* 53, 1549–1552, 1562.

Hibma, A. M., Jassim, S. A., Griffiths, M. W. (1997). Infection and removal of L-forms of *Listeria monocytogenes* with bred bacteriophage. *Int. J.Food Microbiol.*, 4, 197–207.

Hinton, A. R., Trinh, K. T., Brooks, J. D., Manderson, G. J. (2002). Thermophile survival in milk fouling and on stainless steel during cleaning. *Transactions of the Institute of Chemical Engineers*, 80 (Part C), 299–304.

Hoffman, A.D., Gall, K.L., Norton, D.M., Wiedmann, M. (2003). *Listeria monocytogenes* contamination patterns for the smoked fish processing environment and for raw fish. *J. Food Prot.* 66, 52-60.

Holah, J. T. (1992). Industrial monitoring: hygiene in food processing. In: Melo, L. F., Bott, T.R., Hood, S. K., Zottola, E. A. (1997). Growth media and surface conditioning influence the adherence of *Pseudomonas fragi, Salmonella typhimurium* and *Listeria monocytogenes* cells to stainless steel. *J. Food Prot.*, 60, 1034–103.

Holah J. T., Bird, J., Hall, K. E. (2004). The microbial ecology of high-risk, chilled food factories; evidence for persistent *Listeria* spp. and *Escherichia coli* strains. *J. Appl. Microbiol.*, 97, 68–77.

Houry, A., Gohar, M., Deschamps, J., Tischenko, E., Aymerich, S., Gruss, A., Briandet, R. (2012). Bacterial swimmers that infiltrate and take over the biofilm matrix. *Proc. Natl. Acad. Sci. USA*. 109, 13088-13093.

Hung, T. T., Schneewind, O. (2004). Assembly of pili in Gram-positive bacteria. *Trends in Microbiology*, 12, 228-234.

Jamali, H., Radmehr, B., & Thong, K. L. (2013). Prevalence, characterization, and antimicrobial resistance of Listeria species and *Listeria monocytogenes* isolates from raw milk in farm bulk tanks. *Food Control*, 34(1), 121-125.

Jemmi, T,, Keusch, A. (1992). Behavior of *Listeria monocytogenes* during processing and storage of experimentally contaminated hot-smoked trout. *Int. J. Food Microbiol.* 15:339-46.

Jeong, D. K., Frank, J. F. (1994a). , Growth of *Listeria monocytogenes* at 21°C in biofilms with microorganisms isolated from meat and dairy environments. *Lebensmittel-Wissenschaft und-Technologie*, 27, 415–424.

Jeong, D. K., Frank, J. F. (1994b) Growth of *Listeria monocytogenes* at 10 degrees C in biofilms with microorganisms isolated from meat and dairy processing environments. *J. Food Prot.*, 57, 576-586.

Joaquin, J.C., Kwan, C., Abramzon, N., Vandervoort, K., Brelles-Marino, G. (2009). Is gas-discharge plasma a new solution to the old problem of biofilm inactivation? *Microbiol.* 155, 724-732.

Johansen, C., Falholt, P., Gram, L. (1997). Enzymatic removal and disinfection of bacterial biofilms. *Appl. Environ.Microbiol.* 9, 3724–3728.

Johansson, T., Rantala, L., Palmu, L., Honkanen-Buzalski, T. (1999). Occurrence and typing of *Listeria monocytogenes* strains in retail vacuum-packed fish products and in a production plant. *Int. J. Food Microbiol.* 47, 111-119.

Jones, G. W., Isaacson, R. E. (1983). Proteinaceous bacterial adhesins and their receptors. *CRC Critical Rev.Microbiol.*, 10, 229–260.

Joseph, B, Otta, S. K., Karunasagar, I., Karunnasagar, I. (2001). Biofilm formation by *Salmonella spp.* on food contact surfaces and their sensitivity to sanitizers. *Int. J. Food Microbiol.*, 64, 367-372.

Junttila, J. R., Niemel€a, S. I., Hirn, J. (1988). Minimum growth temperatures of *Listeria monocytogenes* and non-haemolytic Listeria. *J. Appl. Bact.*, 65, 321-327

Juven, B. J., Pierson, M. D. (1996). Antibacterial effects of hydrogen peroxide and methods for its detection and quantification. *J. of Food Prot.*, 59, 1233–1241.

Kalmokoff, M. L., Austin, J. W., Wan, X-D., Sanders, G., Banerjee, S., Farber, J. M. (2001). Adsorption, attachment and biofilm formation among isolates of *Listeria monocytogenes* using model conditions. *J.Appl.Microbiol.*, 91, 725-734.

Kamgang, J.O., Briandet, R., Herry, J.M., Brisset, J.L., Naitali, M. (2007). Destruction of planktonic, adherent and biofilm cells of *Staphylococcus epidermidis* using a gliding discharge in humid air. *J. Appl. Microbiol.* 103, 621-628.

Kamgang-Youbi, G., Herry, J.M., Meylheuc, T., Brisset, J.L., Bellon-Fontaine, M.N., Doubla, A., Naitali, M. (2009). Microbial inactivation using plasma-activated water obtained by gliding electric discharges. *Lett. Appl. Microbiol.* 48, 13-18.

Koch, J., Dworak, R., Prager, R., Becker, B., Brockmann, S., Wicke, A., Wichmann-Schauer, H., Hof, H., Werber, D., Stark, K. (2010). Large listeriosis outbreak linked to cheese made from pasteurized milk, Germany, 2006–2007. *Foodborne Path.. Dis.* 7, 1581–1584.

Kudva, I. T., Jelacic, S., Tarr, P. I., Youderian, P., Hovde, C. J. (1999). Biocontrol of *Escherichia coli* O157 with O157-specific bacteriophages. *Applied and Environmetal Microbiology*, 65, 3767.

Harvey, J., Keenan K. P., Gilmour, A. (2007). Assessing biofilm formation by *Listeria monocytogenes* strains. *Food Microbiol.* 24:380–392

Ibba, M., Cossu, F., Spanu, V., Virdis, S., Spanu, C., Scarano, C. De Santis, E.P.L. (2013). *Listeria monocytogenes* contamination in dairy plants: evaluation of *Listeria monocytogenes* environmental contamination in two cheese-making plants using sheeps milk. *It. J. Food Safety*, 2, 109-112.

Latorre, A. A., Van Kessel, J. A., Karns, J. S., Zurakowski, M. J., Pradhan, A. K., Zadoks, R. N. (2009). Molecular ecology of *Listeria monocytogenes*: evidence for a reservoir in milking equipment on a dairy farm. *Appl. and Environ. Microbiol.*, 75, 1315-1323.

Latorre, A.A., Van Kessel, J.S., Karns, J.S., Zurakowski, M.J., Pradhan, A.K., Boor, K.J., Jayarao, B.M., Houser, B.A., Daugherty, C.S., Schukken, Y.H. (2010). Biofilm in milking equipment on a dairy farm as a potential source of bulk tank milk contamination with *Listeria monocytogenes*. *J. of Dairy* Sci. 93, 2792–2802.

Lee, S.H., Frank, J.F., (1991). Inactivation of surface-adherent *Listeria monocytogenes*. Hypochlorite and heat. *J. Food Prot.* 54, 4–6.

Liu, D. (2006). Identification, subtyping and virulence determination of *Listeria monocytogenes*, an important foodborne pathogen. *J. Med. Microbiol.*, 55(Pt 6), 645-659.

López V., Villatoro D., Ortiz S., López P., Navas J., Dávila J.C., Martinéz-Suárez J.V. (2008): Molecular tracking of *Listeria monocytogenes* in an Iberian pig abattoir and processing plant. *Meat Science*, 78: 130–134.

Lu T.K. and J.J. Collins (2007). Dispersing biofilms with engineered enzymatic bacteriophage, *Proceedings of the National Academy of Sciences USA* 104, 11197–11202.

Lunden J.M., Autio T.J., Korkeala, H.J. (2002). Transfer of persistent *Listeria monocytogenes* contamination between food-processing plants associated with a dicing machine. *J. Food Prot.*, 65, 1129–1133.

Mackiw, E., Modzelewska, M., Maka L., Sciezynska, H., Pawłowska, K., Postupolski, J., Korsak, D. (2016). Antimicrobial resistance profiles of *Listeria monocytogenes* isolated from ready-to-eat products in Poland in 2007-2011. *Food Control* 59, 7-11.

Mafu, A.A., Roy, D., Goulet, J. and Hagny, P., 1990. , Attachment of *Listeria monocytogenes* to stainless steel, glass, polypropylene and rubber surfaces after short contact times. *J. Food Prot.* 53, 742–746.

Mahdavi M., M. Jalali and R.K. Kermanshahi, (2007). The effect of nisin on biofilm forming foodborne bacteria using microtiter plate method, *Research in Pharmaceutical Sciences* 2, pp. 113–118.

Markkula, A., Autio, T., Lunden, J., Korkeala, H. (2005). Raw and processed fish show identical *Listeria monocytogenes* genotypes with pulsed-field gel electrophoresis. *J. Food Prot.* 68, 1228-1231

Mashall, K. (1994). Microbial adhesion in biotechnological processes. *Current Opinion in Biotechnology*, 5, 296–301.

Mead, P.S., Slutsker, L., Dietz, V., McCaig, L.F., Bresee, J.S., Shapiro, C., Griffin, P.M., Tauxe, R.V. (1999). Food related illness and death in United States. *Emerg. Infect. Dis.* 5: 607-625.

Medrala, D., Dabrowski, W., Czekajlo-Kolodziej, U., Daczkowska-Kozon, E. Koronkiewicz, A., Augustynowicz, E., Manzano, M. (2003). The possible effect of a sanitization program on intraspecies differentiation of *Listeria monocytogenes* strains isolated from a fish processing plant. *Int. J. Hyg. Environ. Health* 206, 583-590.

Meloni, D.,Galluzzo, P.,Mureddu, A.,Piras, F.,Griffiths, M., Mazzette, R. (2009). *Listeria monocytogenes* in RTE foods marketed in Italy: Prevalence and automated EcoRI ribotyping of the isolates. *Int. J. Food Microbiol.*, 129, 166–173.

Meloni, D. (2014). Focusing on the main morphological and physiological characteristics of the food-borne pathogen *Listeria monocytogenes*. *J. Vet. Sci. Res.* 2014, 1, 1–2.

Meylheuc T., M. Renault and M.N. Bellon-Fontaine, Adsorption of a biosurfactant on surfaces to enhance the disinfection of surfaces contaminated with *Listeria monocytogenes*, *Int.J.Food Microbiol.* 109, 71–78.

Midelet, G., Kobilinsky, A., & Carpentier, B. (2006). Construction and analysis of fractional multifactorial designs to study attachment strength and transfer of *Listeria monocytogenes* from pure or mixed biofilms after contact with a solid model food. *Appl.Environ.Microbiol.*, 72(4), 2313–2321.

Miettinen, H., Wirtanen, G. (2006). Ecology of *Listeria* spp. in a fish farm and molecular typing of *Listeria monocytogenes* from fish farming and processing companies. *Int. J. Food Microbiol.* 112, 138e146.

Miya, S., Takahashi, H,, Ishikawa, T., Fujii, T., Kimura, B. (2010). Risk of *Listeria monocytogenes* Contamination of Raw Ready-To-Eat Seafood Products Available at Retail Outlets in Japan. *Appl.Environm.Microbiol.*, 3383–3386.

Ming, X.T., Weber, G.H., Ayres, J.W. and Sandine, W.E., 1997. , Bacteriocins applied to food packaging materials to inhibit *Listeria monocytogenes* on meats. *J. Food Sci.* 62, pp. 413–415.

Moens, S., and J. Vanderleyden. 1996. Functions of bacterial flagella. *Crit. Rev. Microbiol.* 22:67-100.

Molin, S., Tolker-Nielsen, T. (2003). Gene transfer occurs with enhanced efficiency in biofilms and induces enhanced stabilisation of the biofilm structure. *Current Opinion in Biotechnology*, 14(3), 255–261.

Moltz, A. G., Martin. S. E. (2005). Formation of biofilms by *Listeria monocytogenes* under various growth conditions. J.Food Prot., 68, 92-97.

Mosteller, T.M. & Bishop, J.R. 1993 Sanitizer efficacy against attached bacteria in a milk biofilm. *J. Food Prot.*, 54, 34 -41.

Murphy, S. C., Boor, K. J. (2000). Trouble-shooting sources and causes of high bacteria counts in raw milk. *Dairy Food Environ. Sanit.* 20:606–611

Nilsson, L., Huss, H.H., Gram, L. (1997). Inhibition of *Listeria monocytogenes* on cold-smoked salmon by nisin and carbon dioxide atmosphere. *Int. J. Food Microbiol.*38, 217–228.

Nikolaev Y.A., Plakunov, V.K. (2007). Biofilm -"City of Microbes" or an Analogue of Multicellular Organisms? *Microbiology*. 76 (2), 125-138.

Norton, D.M., McCamey, M.A., Gall, K.L., Scarlett, J.M., Boor, K.J., Wiedmann, M. (2001). Molecular studies on the ecology of *Listeria monocytogenes* in the smoked fish processing industry. *Appl. Environ. Microbiol.* 67, 198-205.

Norwood, D. E., Gilmour, A. (1999). Adherence of *Listeria monocytogenes* strains to stainless steel coupons. *J. Appl. Microbiol.* 86:576–582.

Norwood, D., Gilmour, A. (2001). The differential adherence capabilities of two *Listeria monocytogenes* strains in monoculture and multispecies biofilms as a function of temperature. *Lett. in Appl. Microbiol.* 33, 320–324.

Olsen, J. E. (2004). Role of flagellin and the two-component CheA/CheY system of *Listeria monocytogenes* in host cell invasion and virulence. *Infection and Immunity,* 72, 3237-3244.

O'Toole, G. A., and R. Kolter (1998). Initiation of biofilm formation in *Pseudomonas fluorescens* WCS365 proceeds via multiple, convergent signalling pathways: a genetic analysis. *Mol. Microbiol.* 28:449-461.

Oh, D.H., Marshall, D.L., 1996. , Monolaurin and acetic acid inactivation of *Listeria monocytogenes* attached to stainless steel. *J. Food Prot.* 59, pp. 249–252.

Oliveira M.M., Brugnera D.F., Alves E., Piccoli R.H. (2010). Biofilm formation by *Listeria monocytogenes* on stainless steel surface and biotransfer potential. *Braz. J. Microbiol.* 41, 97-106.

Palmer J., Flint S., Brooks J. (2007). Bacterial cell attachment, the beginning of a biofilm. *J. Ind.Microbiol.Biotech.*, 34, 577–588.

Pan, Y., Breidt Jr., F., Kathariou, S. (2006). Resistance of *Listeria monocytogenes* biofilms to sanitizing agents in a simulated food processing environment. *Appl. Environ. Microbiol.* 72, 7711–7717.

Parkar S.G., S.H. Flint and J.D. Brooks (2004). Evaluation of the effect of cleaning regimes on biofilms of thermophilic bacilli on stainless steel, *J. of Appl.Microbiol.*, 96, 110–116.

Perez-Conesa, D., McLandsborough, L., & Weiss, J. (2006). Inhibition and inactivation of *Listeria monocytogenes* and *Escherichia coli* O157:H7 colony biofilms by micellarencapsulated eugenol and carvacrol. *J. Food Prot.,* 69(12), 2947–2954.

Piana, A., Are, R., Orrù, M., Saba, F., Dettori, M., Maida, I., Sotgiu, G., Rais, C., Mura, M.S (2005). *Listeria monocytogenes* meningoencephalitis: Molecular methods for diagnosis and for monitoring the response to chemotherapy. *Ital. J. Public Health*, 2, 29–34.

Poncin-Epaillard, F., Herry, J.M., Marmey, P., Legeay, G., Debarnot, D., Bellon- Fontaine, M.N. (2013). Elaboration of highly hydrophobic polymeric surfaceea potential strategy to reduce the adhesion of pathogenic bacteria? *Mater. Sci. Eng. C Mater. Biol. Appl.* 33, 1152-1161.

Rodriguez A., Autio W.R., MCLandsborough L.A. (2007). Effects of inoculation level, material hydration, and stainless steel surface roughness on the transfer of *Listeria monocytogenes* from inoculated bologna to stainless steel and high-density polyethylene. *J. of Food Prot.* 70, 1423-1428.

Ronner, A.B.; Wong, A.C.L. (1993). Biofilm development and sanitizer inactivation of *Listeria monocytogenes* and *Salmonella typhimurium* on stainless-steel and buna-N rubber. *J. Food Prot.* 56 (9), 750- 758.

Rorvik, L.M., Caugant, D.A., Yndestad, M. (1995). Contamination pattern of *Listeria monocytogenes* and other *Listeria* spp. in a salmon slaughterhouse and smoked salmon processing plant. *Int. J. Food Microbiol.* 25, 19-27.

Rorvik, L.M. (2000). *Listeria monocytogenes* in smoked salmon industry. Int. *J. Food Microbiol.* 62, 183–190.

Rosmaninho R., O. Santos, T. Nylander, M. Paulsson, H. Müller-Steinhagen and L. Melo, Modified stainless steel surfaces targeted to reduce fouling – evaluation of fouling by milk components, *J. Food Engineering* 80 (2007), pp. 1176–1187.

Rotariu, O., Thomas, D. J. I., Goodburn, K. E., Hutchison, M. L., & Strachan, N. J. (2014). Smoked salmon industry practices and their association with *Listeria monocytogenes. Food Control*, 35, 284-292.

Ryu, J.H., Beuchat, L.R., (2005). Biofilm formation by *Escherichia coli* O157:H7 on stainless steel: effect of exopolysaccharide and Curli production on its resistance to chlorine. *Appl. Environ. Microbiol.* 71, 247e254.

Sandasi M., Leonard, C.M.,Viljoen, A.M. (2008). The effect of five common essential oil components on *Listeria monocytogenes* biofilms, *Food Control* 19, 1070–1075.

Sasahara, K.C. and Zottola, E.A. (1993) Biofilm formation by *Listeria monocytogenes* utilizes a primary colonizing microorganism in fowing systems. *J.Food Prot.* 56, 1022-1028.

Sauer, K., Camper, A.K., Ehrlich, G.D., Costerton, J.W., Davies, D.G., (2002). *Pseudomonas aeruginosa* displays multiple phenotypes during development as a biofilm. *J. Bacteriol.* 184, 1140e1154.

Sharma M. , Anand S.K. (2002). Biofilms evaluation as an essential component of HACCP for food/dairy processing industry – a case. *Food Control,* 13, 469–477.

Sidhu M.S., Langsrud, S., Holck, A. (2001). Disinfectant and antibiotic resistance of lactic acid bacteria isolated from food industry, Microbial *Drug Resistance* 7 (2001), pp. 73–83.

Sillankorva S., D.R. Oliveira, M.J. Vieira, I.W. Sutherland and J. Azeredo (2004). Bacteriophage Φ S1 infection of *Pseudomonas fluorescens* planktonic cells versus biofilms, *Biofouling* 20, 133–138.

Simões M.. Vieira M.J, (2009). Persister cells in *Pseudomonas fluorescens* biofilms treated with a biocide, Proceedings of the international conference processes in biofilms: *Fundamentals to applications*, 58–62 Davis, CA, USA.

Simões M., Simoes L.C., Vieira M.J. (2010). A review of current and emergent biofilm control strategies. *LWT - Food Sci. Tech.*, 43, 573-583.

Sinde E., Carballo J. (2000). Attachment of *Salmonella spp* and *Listeria monocytogenes* to stainless steel, rubber and polytetrafluorethylene: the influence of free energy and the effect of commercial sanitizers, *Food Microbiol.* 17: 439-447.

Singh P.K., M.R. Parsek, P.E. Greenberg and M.J. Welsh, (2002). A component of innate immunity prevents bacterial biofilm development, *Nature* 417, 552–555.

Smith L., Fratamico P.M. , Novak, J.S. (2004) Quorum sensing: a primer for food microbiologists, *J. Food Prot.*, 67, 1053–1070.

Somers, E. B., Wong A. C. L. (2004). Efficacy of two cleaning and sanitizing combinations on *Listeria monocytogenes* biofilms formed at low temperature on a variety of materials in the presence of ready-to-eat meat residue. *J. Food. Protect.* 67:2218-2229.

Speranza B., Sinigaglia M., Corbo M.R., (2009). Non starter lactic acid bacteria biofilms: A means to control the growth of *Listeria monocytogenes* in soft cheese. *Food Control* 20, 1063-1067.

Splendiani A., A.G. Livingston, Nicolella, C. (2006). Control membrane-attached biofilms using surfactants, *Biotechnology and Bioengineering* 94, 15–23.

Stepanovic S, Cirkovic I, Ranin L, Svabic-Vlahovic M. (2004). Biofilm formation by *Salmonella spp.* and *Listeria monocytogenes* on plastic surface. *Lett. Appl. Microbiol.* 38: 428-432.

Sutherland I.W., Hughes K.A., Skillman L.C., Tait K. (2004). The interaction of phage and biofilms. *FEMS Microbiology Letters*, 232, 1–6.

Thimothe, J., Nightingale, K.K., Gall, K., Scott, V.N., Wiedmann, M. (2004). Tracking of *Listeria monocytogenes* in smoked fish processing plants. *J. Food Prot.* 67, 328–341.

Tompkin, R.B., V.N. Scott, D. T. Bernard, W.H. Sveum, and K.S. Gombas (1999). Guidelines to prevent post-processing contamination from *Listeria monocytogenes, Dairy, Food Environ. Sanit.* 19: 551-562.

Tompkin, R.B., (2002). Control of *Listeria monocytogenes* in the food-processing environment. *J. Food Prot.* 65, 709–725.

Unnerstad, H., Bannerman, E., Bille, J., Danielsson-Tham, M., Waak, E., Tham, W. (1996). Prolonged contamination of a dairy with *Listeria monocytogenes. Neth. Milk Dairy J.* 50, 493–499.

Vatanioopaisarn, S., A. Nazli, C. R. E. Dodd, C. E. D. Rees, Waites, W. M. (2000). Effect of flagella on initial attachment of *Listeria monocytogenes* to stainless steel. *Appl. Environ. Microbiol.* 66:860-863.

Vitas, A.I., Aguado, V., Garcia-Jalon, I. (2004). Occurrence of *Listeria monocytogenes* in fresh and processed foods in Navarra (Spain). *Int. J. Food Microbiol.* 90:349-6.

Vogel, B.F., Jorgensen, L.V., Ojeniyi, B., Huss, H.H., Gram, L., (2001a). Diversity of *Listeria monocytogenes* isolates from cold-smoked salmon produced in different smokehouses as assessed by Random Amplified Polymorphic DNA analyses. *Int. J. Food Microbiol.* 65, 83-92

Vogel, B.F., Hansen, L., Truelstrup-Mordhorst, H., Gram, L., (2010b). The survival of *Listeria monocytogenes* during long term desiccation is facilitated by sodium chloride and organic material. *Int. J. Food Microbiol.* 140, 192–200.

Waak, E., Tham, W., Danielsson-Tham, M.L. (2002). Prevalence and fingerprinting of *Listeria monocytogenes* strains isolated from raw whole milk in farm bulk tanks and in dairy plant receiving tanks. *Appl. Environ.* Microbiol. 68:3366–3370.

Walker, S. J., Archer, P., Banks, J. G. (1990). Growth of *Listeria monocytogenes* at refrigeration temperatures. *J. of Appl. Bacteriology*, 68, 157-162.

Whitehead K.A., J. Collingon and J. Verran, (2005). Retention of microbial cells in substratum surface features of micrometer and sub-micrometer dimensions, *Colloids and Surfaces B: Biointerfaces* 41, 129–138.

Wong A.C.L. (1998). Biofilms in food processing environments, *J. Dairy Sci.* 81, 2765–2770.

Zhang, Y., Yeha, E., Hallb, G., Cripeb, J., Bhagwatc, A.A. Menget, J. (2007). Characterization of *Listeria monocytogenes* isolated from retail foods. *Int. J.Food Microbiol.*113, 47-53.

Zhao, T., M. P. Doyle, Zhao P. (2004). Control of *Listeria monocytogenes* in a biofilm by competitive-exclusion microorganisms. *Appl. Environ. Microbiol.* 70:3996-4003.

Zobell C.E., (1943). The effect of solid surfaces upon bacterial activity. *J. of Bacteriol.* 46, 39–56.

Zottola E.A. (1994). Microbial attachment and biofilm formation: A new problem for the food industry? *Food Technol.* 48, 107–114.

Zottola, E. A., Sasahara K. C. (1994). Microbial biofilms in the food processing industry—Should they be a concern? *Int. J. Food Microbiol.* 23:125–148.

In: Listeria monocytogenes
Editor: Tiffany Vicario

ISBN: 978-1-63483-765-1

Chapter 3

LISTERIA MONOCYTOGENES OUTBREAKS: EPIDEMIOLOGICAL UPDATE AND CONTROL POSSIBILITIES

Pellicer Karina*[*] *and Brusa Victoria
Laboratorio de Microbiología de los Alimentos de la Facultad de Ciencias Veterinarias, Universidad Nacional de La Plata, La Plata, Buenos Aires, República Argentina

ABSTRACT

Listeria monocytogenes is a gram positive bacterium which causes illness in animals and humans. This foodborne pathogen is the etiologic agent of listeriosis and can lead to spontaneous abortion in pregnant women, meningitis and encephalitis in newborns, children and immuno-suppressed adults, and gastroenteritis in health persons (Donelly 1994; Pilchová et al. 2014; Posfay-Barbeand Wald 2004; Seeliger 1961; Shih-Yu et al. 2007; Smith et al. 2009). *L. monocytogenes* is frequently detected in environmental samples and can survive for long periods in soil, water, feed and feces, providing be a source of contamination for the food and water chain (CDC (Center for Disease Control and Prevention); Fox et al. 2009).

L. monocytogenes and *L. innocua* are frequently isolated from food processing plants. However, most of the cases of listeriosis transmitted through food are due to the contamination of raw or cooked foods with *L. monocytogenes*. The minimum infective dose of *L. monocytogenes* has not been established yet, although it has been indicated that the intake of up to 100 cells does not affect healthy consumers. The long incubation time of listeriosis makes difficult to identify the pathogen and trace the contaminated food (Gandhi and Chikindas 2007).

There are 13 serovars known of *L. monocytogenes*, and four of them (1/2a, 1/2b, 1/2c y 4b) are responsible for over 95% of reported human listeriosis cases. The major outbreaks have been caused by *L. monocytogenes* 4b. In general, mortality rate is low, however in inmuno-supressed and susceptible patients this rate is 20 – 30% (Donelly 1994).

---

[*] Corresponding author: pellicerk@fcv.unlp.edu.ar.

Incidence

Listeria is widespread in the environment; therefore, a wide range of different foodstuffs can be contaminated. Also, isolation from animals is to be expected and increased exposure may lead to clinical disease in animals (EFSA (European Food Safety Authority), 2013).

L. monocytogenes has been isolated from various food sources, such as dairy products, fresh vegetables and meats (Beresford, Andrew, and Shama 2001; Guerrieri et al. 2009). Listeriosis has become a serious public health problem and several outbreaks in European countries, mainly related with dairy product consumption, have been produced (Goulet, Hedberg, Le Monnier and De Valk 2008; Little, Sagoo, Gillespie, Grant and McLauchlin 2009; Lundén, Tolvanen and Korkeala 2004). According to EFSA, the annual incidence of listeriosis varies between 0.3 and 7.5 cases per one million people. EFSA reported 1.470 human cases with a mortality rate of 12.7%, in EU (European Union) during 2011.In 2013 1.763 human cases of listeriosis (notification rate 0.44 cases per 100.000 populations) were confirmed, showing an increase of 8.6%, compared with 2012.A total of 191 deaths due to listeriosis were reported in 2013 in the EU. In the same year, a total of twelve listeriosis outbreaks were reported by seven Member States (MS) of EU. Seven of the outbreaks reported were supported by strong evidence. Crustaceans, shellfish, mollusks and sub-products were implicated in three strong-evidence outbreaks. In two of these outbreaks, the source was crab meat. The responsible food vehicles in the remaining four outbreaks belonged to four different food categories ('Cheese', 'Meat and meat products', 'Pig meat and sub-products', 'Vegetables and juices and other sub-products'). For a healthy human population, foods not exceeding the level of 100 CFU/g are considered to pose a negligible risk. Therefore, the EU microbiological criterion for *L. monocytogenes* in Ready-To-Eat (RTE) food is set at ≤ 100 CFU/g for RTE products on the market. As in previous years and consistent with the results of the EU baseline survey on the prevalence of *L. monocytogenes* in certain RTE foods at retail (EFSA 2013a), the proportion of positive samples at retail was highest in fish products (mainly smoked fish), followed by soft and semi-soft cheeses, RTE meat products and hard cheeses. Several MS reported findings of Listeria in animals. Most of the tested samples were from cattle, and to a lesser degree goats and sheep. Findings of Listeria were most often reported in these three animal species, but were also detected in fowl, pigs, dogs, foxes, horses, African wild dogs and alpacas.

In accordance with the CDC of the United State (US), every year around of 1.600 people get Listeriosis, and 260 die by this cause in US (Scallan et al. 2011). At least 90% of patients are in a higher risk group. Healthy children and adults occasionally get infected with Listeria, but they rarely become seriously ill. In 2012, 831 foodborne outbreaks were reported to CDC (CDC, 2014a). In 2013, the average annual incidence of this foodborne in the United States was 0.26 cases per 100.000 individuals (CDC, 2014b). During 2013 and 2014, 735 and 769 listeriosis human cases were confirmed, respectively.

Growth Behavior and Control

L. monocytogenes can survive and grow at different of environmental conditions; due to possess a wide range of physiological response mechanisms for adaptation against diverse stresses conditions.Refrigeration temperatures, low pH and high salt concentration allow the pathogen to overcome food preservation and safety barriers, and pose a potential risk to human health (Gandhi and Chikindas 2007).

Due to its psychotropic profile, *L. monocytogenes* can survival down to a few degrees below 0°C. Also, *L. monocytogenes* has the ability to grow over a wide range of temperatures (2–45°C), and can survive and growth at refrigeration temperatures. The changes that occur in carbon and fatty acids membrane composition of *L. monocytogenes*, helps maintain the degree of membrane fluidity for growth, when the temperature is reduced below optimum. The changes include an increase in the proportion of C15:0, a modification on the degree of unsaturated fatty acids, and the shortening of fatty acid chain length that decreases the carbon–carbon interaction between neighboring chains in the cell membrane (Beales 2004).

In response to a temperature shock, changes in *L. monocytogenes* gene expression are produced. Cold shock proteins (Csps) and cold acclimation proteins (Caps) and synthesize during balanced growth at low temperatures (Bayles, Annous and Wilkinson 1996). Other strategy of *L. monocytogenes* is to accumulate solutes cryoprotectants such as glycine betaine and carnitine, and the role of these compounds as cryoprotectants have been widely studied. Wemekamp-Kamphuis et al. (2004) demonstrated that deletions of these osmolyte transporters reduced the growth of Listeria at low temperatures (WemekampKamphuis et al. 2004). The mechanisms described for *L. monocytogenes* survival and growth at low temperatures, shown the versatility of this emerging pathogen to adapt to a wide range of environmental conditions.

L. monocytogenes can grow between pH 4.6 and 9.5 (Gandhi and Chikindas 2007), and while the acid pH may be inhibitory or lethal to *L. monocytogenes*, have been demonstrated that some strains tolerate pH 4.4. *L. monocytogenes* responds to and survives in these low-pH environments by utilizing a number of stress adaptation mechanisms. Exposure of *L. monocytogenes* to mild acidic pH of 5.5 (1M lactic acid) induces the acid tolerance response (ATR), wherein the cells are resistant to severe acidic conditions (O'Driscoll, Gahan and Hill 1996). Different proteins are synthesized by exposing cells to a lethal and a non-lethal acidic pH (acid stress and acid adaptation). Although, more proteins were induced in cells exposed to the lethal pH, a majority of the proteins induced were common to both pH conditions used (Phan-Thanh and Mahouin 1999).In addition, acid-adapted *L. monocytogenes* increased resistance to heat shock (52°C), osmotic shock (25–30% NaCl) and alcohol stress, suggesting that acid adaptation also provides cross-protection against other stress factors (Phan-Thanh et al. 2000).

In order to survival under acid stress, *L. monocytogenes* utilizes the glutamate decarboxylase (GAD) system to survive. The GAD system is composed of three genes *gadA*, *gadB* and *gadC* genes. It has been proposed that glutamate is taken up by the cell via a specific transporter followed by its decarboxylation in the cytoplasm, producing γ-aminobutyrate and resulting in the utilization of an intracellular proton. The γ-aminobutyrate is then exported from the cell via an antiporter located in the cell membrane. The proton loss from the cell results in an increase in the pH of the cytoplasm and the release of alkaline γ-aminobutyrate into the environment raises the external pH slightly (Waterman and Small 1998). The functional GAD system is vital for the acid resistance of *L. monocytogenes* and to successfully pass through the gastric environment and infect the small intestine (Cotter Gahan and Hill 2001).

This bacteria survives on foods with water activity as low as 0.92 (Gandhi and Chikindas 2007) and with a high concentration of sodium chloride (20% NaCI). The survival under osmotic stress involves both physiological changes and variations of gene expression patterns and is called osmoadaptation (Hill, Cotter, Sleator and Gahan 2002). One of the adapting mechanisms used by Listeria to salt stress is a change in gene expression, leading to an increased or decreased synthesis of various proteins. Similar to the two groups of proteins induced in response to cold shock, salt shock proteins (Ssp) and the stress acclimation proteins (Sap) were identified (Duchè, Trèmoult, Glaser and Labadie 2002a). Ssp and Sap are rapidly induced and continue to be over expressed several hours after conditions return to normal. Among the eleven Saps identified, GbuA, which functions as an osmoprotectant transporter for glycine betaine was induced in response to salt stress. The Ctc protein is aSsp, which is induced in response to salt stress (Duchè, Trèmoult, Glaser and Labadie 2002a, b). This protein is involved in the resistance of *L. monocytogenes* to high osmolarity in the absence of osmoprotectants, such as glycine betaine and carnitine in the medium (Gardan, Duchè, Leroy-Sètrin, Labadie and European Listeria Genome Consortium, 2003).Glycine and prolinebetaine, acetyl carnitine, carnitine, γ-butyrobetaine and 3-dimethylsulphoniopropionate are recognized as osmoprotectants in *L. monocytogenes* (Bayles and Wilkinson, 2000). The presence of these compounds resulted in an up to 2.6-fold increase in growth rate of salt-stressed cells compared to stress cells without any osmoprotectants. The cells take up osmolytes from the external environment as a response to osmotic stress, which helps to regain the osmotic balance within cells (Bayles and Wilkinson, 2000).

It has been reported that the adaptation to high osmotic stress requires expression of two genes, *kdpE* and *orf X*, and depends on the potassium level in the medium (Brøndsted, Kallipolitis, Ingmer and Knöchel, 2003; Kallipolitis and Ingmer, 2001). Thus, the uptake of potassium from the environment via the Kdp system has a protective effect on *L. monocytogenes* against salt stress.

L. monocytogenes cells can be fixed onto various surfaces, especially in inaccessible parts of industrial equipment due to biofilm formation. This can occur rapidly and the substrate, once attached, is difficult to remove. *L. monocytogenes* has been found in biofilms on plastic surfaces, polypropylene, rubber, stainless steel and glass. In previous work (Copes, Pellicer, Malvestiti and Stanchi, 2000) we have observed that *Listeria* spp. can survive on plastic and wood food cutting boards until 75 days. The ability of *L. monocytogenes* to form biofilms provide protection to the microorganisms in food processing environments (Ho, Lappi and Wiedmann 2007) and equipment (Fox, Hunt, O'Brien and Jordan 2011). The biofilms prevent the accessibility of sanitizers and disinfectants (Arizcun, Vasseur and Labadie 1998; Simões, Simões and Vieira 2010), allowing the persistence of this bacteria on food-processing surfaces. Also, contributes to the risk (Aarnisalo, Raaska and Wirtanen 2007) of cross- or re-contamination of food products (Beresford, Andrew and Shama, 2001; Renier, Hébraud and Desvaux, 2011; Tresse et al. 2007). A novel approach to prevent *L. monocytogenes* biofilm formation and to eliminate the pathogens in established biofilms, involves the use of lactic acid bacteria (LAB) and their metabolic products. It has been reported that the undissociated LAB inhibits the growth of microorganisms and is frequently used to extend the shelf life of foods. Lactic acid is effective to inhibit the development of *L. monocytogenes* when it reaches pH values of 4.5 and 4.6. Lactic acid bacteria LAB antagonistic products, such as nisin, reuterin and pediocin, have been well documented for their biofilm control potential against different microorganisms including *L. monocytogenes*. Bacteriocins are increasingly used to control the growth of *L. monocytogenes*; among them, nisin is an antimicrobial agent that can be used in meat, in combination with pH reduction and the addition of NaCl.

Given that *L. monocytogenes* is transmitted by food consumption, mainly RTE products, some strategies are proposed in order to prevent the development of Listeria in those products using the hurdle technology (Leitsner and Gorris 1995; Pellicer et al. 2011). The aim of hurdle technology is to improve the total quality of food by application of a strategic mix of hurdles (Leistner 2000). The physiological responses of microorganisms during food preservation are the basis for the application of advanced hurdle technology. When this technology for food preservation is applied, gentle hurdles used over foodstuffs will have a synergistic effect. The hurdle technology was the basis for the development of predictive microbiology, whose aims are predict the shelf life and safety of foodstuffs, determine the microbiological stability of new food products, and assist in risk assessment of foodstuffs under different environmental conditions.

The ability to grow at refrigeration temperatures is particular important in refrigerated for RTE products. The psychrotrophic characteristic of Listeria, increases the risk factors for RTE foods products (Tiwari, Walsh, Rivas, Jordan and Duffy 2014) and make difficult the control of this foodborne pathogen (Rocourt and Cossart, 1997). Since refrigeration is one of the most common ways to increase the shelf life of foods, understanding the mechanisms behind its survival and growth at low temperature could provide information to help develop more effective control methods for the pathogen (Gandhi and Chikindas 2007).

The cross-resistance of acid adapted cells to other stresses has important implications for the food industry, particularly since foods commonly encounter sublethal acidic treatments during processing (van Schaik, Gahan and Hill 1999). The acidic pH of many foods is one of the many factors that help to prevent the growth of foodborne pathogens. Therefore, the acid tolerance response observed in *L. monocytogenesis* of particular concern during food processing, because an exposure of the pathogen to mild acidic conditions could confer resistance to more severe acidic conditions. As was explained above, lactic acid is used to extend the shelf life of foods, and is effective to inhibit the development of *L. monocytogenes* when at pH values ranged between 4.5 and 4.6 (Ahmad and Marth 1990; Ariyapitipun, Mustapha and Clarke 2000; Sorrels, Engil and Hatfield 1989).

The use of salt to lower the water activity is one of the methods of food preservation used by the food industry; however, the ability of Listeria to adapt and survive in high concentrations of salt makes it difficult to control the pathogen in foods. In addition to individual stress factors, it is important to keep in mind that cross-protection to environmental stresses is commonly seen as part of the stress response of Listeria in foods. This is crucial when deciding on food processing and preservation parameters, since exposure of the pathogen to one kind of sub-lethal stress can confer cross-protection to other lethal stresses.

Contamination of foods can occur during any stage of the manufacturing or processing phase. Despite the difficulty and uncertainty in identifying the source of contamination in foodborne disease outbreaks, several surveillance reports have shown that post-process contamination of foods has been a major cause in many of the outbreaks. To deal with the problem of post-process contamination, research has led to the development of active packaging, wherein materials are incorporated into the packaging to either control the atmosphere within the package (such as moisture content, pH, oxygen level) or inhibit the growth of spoilage and pathogenic organisms on the food product.

Several species of Listeria isolated from humans, food and processing facilities are resistant to one or more antibiotics. In addition to antibiotic resistance, the emergence and spread of resistance among foodborne organisms to sanitizers and disinfectants used by the food industry are also becoming a concern. Regardless of the disinfectant used, it is extremely necessary to control this microorganism, respect strictly to standardized procedures sanitation (cleaning, degreasing, and disinfection) and avoid using materials that promote biofilm formation.

Many studies were done to determine the susceptibility of Listeria to quaternary ammonium compounds (QACs), commonly used as disinfectants in food processing facilities. Bacteriocins are antimicrobial peptides produced by bacteria, such as lactic acid bacteria. These compounds are synthesized from ribosomes and are effective against closely related bacteria (Klaenhammer 1993). Nisin is bacteriocin produced by *Lactococcus lactis* strains. This amino-acid bacteriocin acts on target cells by permeabilizing the cytoplasmic membrane. The formation of pores leads to the leakage of cytoplasmic substances from the cell (Abee, Rombouts, Hugenholtz, Guihard and Letellier 1994). Nisin was approved for use in food preservation in many countries and has been used as a preservative in the dairy and meat industries to control pathogens. The widespread use of antimicrobial agents such as antibiotics, sanitizers or disinfectants in food industry or equipment cleaning, and their effect on antimicrobial resistance is being investigated. Genetic factors such as the mobility of antibiotic resistance genes found on plasmids and transposons can increase the transfer of antibiotic resistance between bacteria. The ability of bacteria to adapt to adverse environmental conditions is an important factor in the development of resistance. This is because an exposure of the organism to a sub-lethal level of an antimicrobial agent can lead to adaptation and development of resistance to higher levels of the antimicrobial or even cross-resistance to other agents.

The growing demand for fresh, minimally processed foods by consumers has led to the need for natural food preservation methods such as the use of antimicrobial peptides to control the growth of foodborne pathogens that have no adverse effects on the consumer or the food itself. The bacteriocinnisin has been widely used as a preservative to control the growth of pathogens in foods. Nisin has also proved very useful as a part of hurdle technology, where a combination of two or more treatments is used to obtain a more effective method of food preservation (Cleveland, Montville, Nes and Chikindas 2001). The combined action of nisin and carbon dioxide on *L. monocytogenes* cells grown at 4 °C has been investigated (Nilsson et al. 2000). Nisin had no effect on nisin-resistant cells grown in the presence of air or carbon dioxide. The presence of carbon dioxide increases the membrane permeability and the proportion of short-chain fatty acids in the cell membrane, which helps in the pore formation by nisin (Nilsson et al. 2000). Also, nisin can be used in meat products (Brandt et al. 2010; Breukink and Krujiff 1999) in combination with pH reduction and the addition of NaCI (Bouttefroy and Millière 2000). Its use is permitted in dairy products (FAO Nutrition Meeting Report Series Nº45 A (1969) but still does not have approval for its application in meat products in South America; some countries such as Japan (Hiroaki Koiso, 2010), Australia and New Zealand had a regulation of its use in meat products (Food Standards Agency of Australia and New Zealand (A565)). In Europe nisin is included in the list of allowed additives (regulation UE Nº 1129/2011) and its use is in meat products were included by Directive 95/2/EC ratified by the Commission of Codex Alimentarius FAO/WHO, 2010 (CX/FA 10/42/5 Add.2 January 2010); while in US is included in the

Listing of Specific Substances Affirmed as GRAS (generally recognized as safe) of Direct Food Substances Affirmed as Generally Recongnized as Safe (Code of Federal Regulations Title 21, Volume 3 in 2014: 21CFR184.1538).

Its potential application is high, especially to avoid the deterioration caused by Gram-positive bacteria in processed meat such as sausage and meat paste in which the homogenization of the product allows the better distribution of nisin (Ariyapitipun, Mustapha and Clarke, 2000; FDA, 1988; Gonalves and Massaguer, 1999; Pellicer et al. 2011). An enhancement of the bactericidal effect of lactic acid with the addition of nisin has been demonstrated.

In previous work has been demonstrated by using the hurdle technology as strategy to prevent the development of Listeria in RTE products, that some combinations are most effective than others (Leitsner and Gorris 1995; Pellicer et al. 2011). The combination of lactic acid with sodium nitrite showed a higher inhibitory index than the combination of lactic acid with nisin (Pellicer et al. 2011).

There has been an ongoing effort to control the foodborne pathogen Listeria in foods and in food processing facilities. Research performed by universities, government agencies and the food industry is aimed at developing new and improved methods to prevent the survival and growth of Listeria. The wide range of efforts to achieve food safety include better monitoring and reporting of foodborne diseases by government agencies, routine food sampling and testing, establishment of HACCP, inspection at food processing facilities, training of food workers and general awareness among consumers about food safety (Bryan, 2002).

In European Union (EU), notification of Listeria in food and animals is required in twelve and thirteen MS, respectively, and the notification of listeriosis in humans is mandatory in most MS (EFSA 2013). Due to the contamination level of foods associated with listeriosis outbreaks, significantly higher than 100 CFU/g (FAO/WHO, 2004), in EU has been determined a maximum level of 100 CFU/g or absence in 25g depending on the RTE product.

A wide spectrum of tolerance mechanisms harbored in *L. monocytogenes*, allows this bacterium to survive under adverse conditions. Due to the ability to growth in stressful medium, including refrigeration temperatures, low pH, high salt concentration, and low activity water; and the capacity of *L. monocytogenes* to biofilm formation; the control of this bacterium is a challenge. Continuous efforts from different areas such as food industry, universities and research groups working hard with this pathogen, maintained the battle for *L. monocytogenes* control in foods. There is no single method that works efficiently when is applied alone. The hurdle technology has demonstrated be the best strategy to try control this pathogen. Therefore a comprehensive plan that includes good manufacturing practices and systematic controls on the production line is mandatory to maintain *L. monocytogenes* charge within allowed values.

References

Aarnisalo, K., Raaska, L. & Wirtanen, G. (2007). Survival and growth of *Listeria monocytogenes* in lubricants used in the food industry. *Food Control.*, *18*, 1019–1025.

Abee, T., Rombouts, F. M., Hugenholtz, J., Guihard, G. &Letellier, L. (1994). Mode of Action of Nisin Z against *Listeria monocytogenes* Scott A Grown at High and Low Temperatures. *Appl. Environ. Microbiol.,60*, 1962–1968.

Ahmad, N. &Marth, E. (1990).Acid-injury of *Listeria monocytogenes.J Food Prot., 53*, 26–29.

Ariyapitipun, T., Mustapha, A., & Clarke, A. D.(2000). Survival of *Listeria monocytogenes* Scott A on vacuum-packaged raw beef treated with polylactic acid, lactic acid, and nisin. *J Food Prot.*, 2000, *63*,131–6.

Arizcun, C., Vasseur, C. & Labadie, J. C. (1998). Effect of Several Decontamination Procedures on *Listeria monocytogenes* Growing in Biofilms. *J Food Prot., 61*,731–734.

Bayles, D. O., Annous, B. A. &Wilkinson, B. J. (1996). Cold stress proteins induced in *Listeria monocytogenes* in response to temperature downshock and growth at low temperatures. *Appl Environ Microbiol., 62*, 1116–1119.

Bayles, D. O. & Wilkinson, B. J. (2000).Osmoprotectants and cryoprotectants for *Listeria monocytogenes. Lett Appl Microbiol., 30*, 23–27.

Beales, N. (2004). Adaptation of microorganisms to cold temperatures, weak acid preservatives, low pH, and osmotic stress: a review. *Compr Rev Food Sci Food Saf., 3*, 1–20.

Beresford, M. R., Andrew, P. W. &Shama, G. (2001). *Listeria monocytogenes* adheres to many materials found in food-processing environments. *J Appl Microbiol., 90*, 1000–1005.

Bouttefroy, A. &Millière, J. B. (2000).Nisin–curvaticin 13 combinations for avoiding the re growth of bacteriocinresistent cells of *Listeria monocytogenes* ATCC 15313.*Int J Food Microbiol., 62*, 65–75

Brandt, A. L., Castillo, A., Harris, K. B., Keeton, J. T., Hardin, M. D. & Taylor, T. M. (2010).Inhibition of *Listeria monocytogenes* by Food Antimicrobials Applied Singly and in Combination. *J Food Sci., 75*, 557–563.

Breukink, E. & de Kruijff, B. (1999).The lantibioticnisin, a special case or not? *Biochim Biophys Acta., 1462*, 223–234.

Bryan, F. L. (2002). Where we are in retail food safety, how we got to where we are, and how do we get there? *J Environ Health., 65*, 29–36.

Brøndsted, L., Kallipolitis, B. H., Ingmer, H. &Knöchel, S. (2003). kdpE and a putative RsbQ homologue contribute to growth of *Listeria monocytogenes* at high osmolarity and low temperature. *FEMS, 219*, 233–239.

CDC. Center for Disease Control and Prevention.http:// www.CDC.gov/listeria.

CDC, (2014a).Center for Disease Control and Prevention.(2014). Surveillance for Foodborne Disease Outbreaks, UnitedStates, 2012, Annual Report. Atlanta, Georgia: US Department of Health and Human Services, *CDC*.

CDC, (2014b).Center for Disease Control and Prevention.(2014). Incidence. *MMWR Morb Mortal Wkly Rep., 63*, 328–332.

Cleveland, J., Montville, TJa., Nes, IF. & Chikindas, M. L., (2001).Bacteriocins: safe, natural antimicrobials for food preservation. *Int J Food Microbiol., 71*, 1-20.

Code of Federal Regulations Title 21, Volume3 in 2014, 21CFR184, 1538.

Commission of Codex Alimentarius FAO/WHO, 2010 (CX/FA 10/42/5 Add.2 Enero de 2010)

Copes, J.,Pellicer, K., Malvestiti, L. & Stanchi, N. O. (2000). Sobrevivencia en tablas de cocina de madera y plástico inoculadas experimentalmente con *Listeria monocytogenes*. *Analecta Vet., 20.*

Cotter, P. D., Gahan, C. G. M. & Hill, C. (2001). A glutamate decarboxylase system protects *Listeria monocytogenes* in gastric fluid. *Mol Microbiol., 40*, 465–475.

Donelly, C. W. (1994). *Listeria monocytogenes*. In: Hui YH, Gorham JR, Murrell KD, Cliver DO(ed.). *Foodborne disease handbook.*, 215–252. Marcel Dekker, New York.

Duché, O., Trémoulet, F., Glaser, P. &Labadie, J. (2002a). Salt Stress Proteins Induced in *Listeria monocytogenes*. *Appl Environ Microbiol., 68*, 1491–1498.

Duché, O.,Trémoulet, F., Namane, A. & Labadie, J. (2002b).A proteomic analysis of the salt stress response of *Listeria monocytogenes*. *FEMS, 215*, 183–188.

European Food Safety Authority and European Centre for Disease Prevention and Control.(2013). The European Union summary report on trends and sources of zoonoses, zoonotic agents and food-borne outbreaks in 2013. EFSA, Parma, Italy.ECDC, Stockholm, Sweden.

FAO (1969). Nutrition Meeting Report Series N° 45 A. www.fao.org.

FAO/WHO. (2004). Risk assessment. *Microbiological Risk Assessment Series, 5.*

FAO/OMS (2010). Directive 95/ 2/ CE:https:// www.fsai.ie/ uploadedFiles/ 95_2_EC.pdf.

FDA.(1988). http:// www.accessdata.fda.gov/ scripts.

Food Standards Agency of Australia and New Zealand (A565). Use of nisin in processed meat product.

Fox., O'Mahony, T., Clancy, M., Dempsey, R., O'Brien, M.& Jordan. (2009). Listeria monocytogenes in the Irish dairy farm environment. *J Food Prot., 72*, 1450–6.

Fox, E., Hunt, K., O'Brien, M. & Jordan, K. (2011). *Listeria monocytogenes* in Irish Farmhouse cheese processing environments. *Int J Food Microbiol., 145*, 39–45.

Gandhi, M. &Chikindas, M. L. (2007). Listeria: A foodborne pathogen that know show to survive. *Int J Food Microbiol., 113*, 1–15.

Gardan, R., Duché, O., Leroy-Sétrin, S., Labadie, J. & European Listeria, Genome Consortium. (2003). Role of ctcfrom Listeria monocytogenes in osmotolerance. *Appl Environ Microbiol., 69*, 154–161.

Gonalves, J. D. &Massaguer, P. R. (1999).The effect of antimicrobials in vacuum-packaged hot dog sausage. IFT´99 Annual Meeting.

Goulet, V., Hedberg, C., Le Monnier, A. & de Valk, H. (2008).Increasing Incidence of Listeriosis in France and Other European Countries. *Emerg Infect Dis., 14*, 734–740.

Guerrieri, E., de Niederhäusern, S., Messi, P., Sabia, C., Iseppi, R., Anacarso, I. &Bondi, M. (2009). Use of lactic acid bacteria (LAB) biofilms for the control of *Listeria monocytogenes* in a small-scale model. *Food Control., 20*, 861–865.

Hill, C., Cotter, P. D., Sleator, R. D. & Gahan, C. G. M. (2002).Bacterial stress response in *Listeria monocytogenes*: jumping the hurdles imposed by minimal processing. *Int Dairy J., 12*, 273–283.

Hiroaki, Koiso. (2010). The use of nisin as a food additive. *Foods & Food Ingredients J. Jpn.*,Vol, *215*, No.4.

Ho, A. J., Lappi, V. R. &Wiedmann, M. (2007). Longitudinal Monitoring of *Listeria monocytogenes* Contamination Patterns in a Farmstead Dairy Processing Facility. *J Dairy Sci., 90*, 2517–2524.

Kallipolitis, R. M. &Higmer, H. (2001) *Listeria monocytogenes* response regulators important for stress tolerance and pathogenesis. *FEMS Microbiol Lett.*, *204*, 111–115.

Klaenhammer, T. R. (1993). Genetics of bacteriocins produced by lactic acid bacteria. *FEMS*, *12*, 39-85.

Leistner, L. (2000). Basic aspects of food preservation by hurdle technology. *Int J Food Microbiol.*, *55*,181–186.

Leistner, L. &Gorris, L. G. M. (1995).Food preservation by hurdle technology. *Trends in Food Science & Technology.*, *6*, 41–46.

Little, C. L., Sagoo, S. K., Gillespie, I. A., Grant, K. &McLauchlin, J. (2009).Prevalence and level of *Listeria monocytogenes* and other Listeria species in selected retail ready-to-eat foods in the United Kingdom. *J Food Prot.*, *72*,1869-1877.

Lundén, J.,Tolvanen, R. & Korkeala, H. (2004). Human Listeriosis Outbreaks Linked to Dairy Products in Europe . *J Dairy Sci.*, *87*, E6–E12

Nilsson, L., Chen, Y.,Chikindas, M. L., Huss, H. H., Gram, L. & Montville, T. H. (2000) Carbon Dioxide and Nisin Act Synergistically on *Listeria monocytogenes. Appl Environ Microbiol.*, *66*, 2769-2774.

O'Driscoll, B., Gahan, C. G. & Hill, C. (1996). Adaptive acid tolerance response in *Listeria monocytogenes*: isolation of an acid-tolerant mutant which demonstrates increased virulence. *Appl Environ Microbiol.*, *62*, 1693–1698

Pellicer, K., Copes, J., Giannuzzi, L. &Zaritzky, N. (2011). Behavior of *Listeria monocytogenes* type1 355/98 (85) in meat emulsions as affected by temperature, pH, water activity, fat and microbial preservatives. *Food Control.*, *22*, 1573–1581.

Phan-Thanh, L. & Mahouin, F. (1999).A proteomic approach to study the acid response in *Listeria monocytogenes. Electrophoresis.*, *20*, 2214–2224.

Phan-Thanh, L., Mahouin, F. & Aligé, S. (2000). Acid responses of *Listeria monocytogenes. Int J Food Microbiol.*, *55*, 121–126.

Pilchováa, T., Hernouldb, Prévost H., Demnerová, K., Pazlarová, J. &Tresse, O. (2014). Influence of food processing environments on structure initiation of static biofilm of *Listeria monocytogenes. Food Control.*, *35*, 366–372.

Posfay-Barbe, K. M. & Wald, E. R. (2004).Listeriosis. *Pediatr Rev.*, *25*, 151–159.

Regulation (EU) No 1129/2011.

Renier, S., Hébraud, M. &Desvaux, M. (2011). Molecular biology of surface colonization by *Listeria monocytogenes*: an additional facet of an opportunistic Gram-positive foodborne pathogen. *Environ Microbiol.*, *13*, 835–850.

Rocourt, J. &Cossart, P. (1997). *Listeria monocytogenes.* In: Doyle MP, Beuchat LR and Montville TJ (ed.). *Food microbiology—fundamentals and frontiers.*, p. 337-352. American Society for Microbiology Press, Washington, D.C.

Scallan, E., Hoekstra, R. M., Angulo, F. J., Tauxe, R. V., Widdowson, M. A., Roy, S. L., Jones, J. L. & Griffin, P. M. (2011). Foodborne illness acquired in the United States--major pathogens. *Emerg Infect Dis.*, *17*, 7–15.

Seeliger, H.P.R. (1961). *Listeriosis. KargerVerlag,Basel.*

Shih-Yu, C., Frank Leigh, L., Ping-Ing, L., Chun-Yi, L., Chien-Yi, C. & Hung-Chieh, C. (2007).Neonatal listeriosis. *J Formos Med Assoc.*, *106*, 161–4.

Simões, M., Simões, L. C. &Vieira, M. J. (2010). A review of current and emergent biofilm control strategies. *LWT - FoodScience and Technology.*, *43*, 573–583.

Smith, B., Kemp, M., Ethelberg, S., Schiellerup, P., Bruun, B. G., Gerner-Smidt, P. & Christensen, J. J. (2009). *Listeria monocytogenes*: Maternal-faetal infections in Denmark 1994-2005. *Scand J Infect Dis.*, *41*, 21–25.

Sorrels, K. M., Engil, D. C. & Hatfield, J. R. (1989). Effect of pH, acidulate, time, and temperature on the growth and survival of *Listeria monocytogenes. J Food Prot.*, *52*, 571–573

Tiwari, U., Walsh, D., Rivas, L., Jordan, K. & Duffy, G. (2014).Modelling the interaction of storage temperature, pH, and water activity on the growth behaviour of *Listeria monocytogenes* in raw and pasteurized semi-soft rind washed milk cheese during storage following ripening. *Food Control.*, *42*, 248–256.

Tresse, O., Shannon, K., Pinon, A., Malle, P., Vialette, M. &Midelet-Bourdin, G. (2007).Variable Adhesion of *Listeria monocytogenes* Isolates from Food-Processing Facilities and Clinical Cases to Inert Surfaces. *J Food Prot.*, *70*, 1569–78.

vanSchaik, W., Gahan, C. G. M. & Hill, C. (1999). Acid-Adapted *Listeria monocytogenes* Displays Enhanced Tolerance against the LantibioticsNisin and Lacticin 3147. *J Food Prot.*, *62*, 536-539.

Waterman, S. R. &Small, P. L. C.(1998).Acid-Sensitive Enteric Pathogens Are Protected from Killing under Extremely Acidic Conditions of pH 2.5 when They Are Inoculated onto Certain Solid Food Sources. *Appl Environ Microbiol.*, *64*, 3882–3886.

Wemekamp-Kamphuis, H. H., Wouters, J. A., de Leeuw, P. P. L. A., Hain, T., Chakraborty, T. &Abee, T. (2004). Identification of Sigma Factor σB-Controlled Genes and Their Impact on Acid Stress, High Hydrostatic Pressure, and Freeze Survival in *Listeria monocytogenes* EGD-e. *Appl Environ Microbiol.*, *70*, 3457–346.

In: Listeria monocytogenes
Editor: Tiffany Vicario

ISBN: 978-1-63483-765-1

Chapter 4

LISTERIA MONOCYTOGENES AND READY-TO-EAT MEAT-BASED FOOD PRODUCTS: INCIDENCE AND CONTROL

A. R. Henriques and M. J. Fraqueza
CIISA, Faculdade de Medicina Veterinária, ULisboa;
Avenida da Universidade Técnica, Lisboa, Portugal

ABSTRACT

Human listeriosis is a serious disease with severe manifestations and long term sequelae in some patients, although the relative low incidence (0.3-0.7 cases/ 100,000 population). Listeriosis data is typically restricted to developed countries, although many of these do not consider the disease as a notifiable one. European member states have reported a statistically significant increasing trend over the period 2009-2013, predominantly related with ready-to-eat foods (RTE) consumption. RTE meat-based food products (RTEMP) are one of the most consumed food products around the world. Their long shelf-lives in refrigerated storage, ease of use, as well as the fact of not requiring any heat-treatment prior to consumption, makes them an important source of food-borne disease, especially by *L. monocytogenes*. This ubiquitous bacterium, widely distributed throughout the natural environment, has a psychrotrophic character and a long persistence capability in the food processing environment due to its ability to form biofilms. These appear to be the main causes of *L. monocytogenes* high prevalence in refrigerated RTEMP. There are several ways by which *L. monocytogenes* can remain in the finished RTEMP product, namely by its recontamination following a listericidal treatment as a result of the contact with contaminated processing equipments or surfaces, in processes such as slicing, cutting, shredding and packaging. Also, this contamination might occur at several stages before consumption, namely in the manufacturing industry, retail establishments, institutional kitchens and even at household level. In retail, significant differences were found in sliced RTEMP depending on the presentation type, having the in-store packaged products a significantly higher incidence than the manufacturer packaged products. So, in this review, available data on *L. monocytogenes* in different RTEMP processing stages along the food chain will be presented. Surveys on *L. monocytogenes* in ready-to-eat foods showed that approximately 2% of RTEMP were contaminated at the end of shelf-life. But since RTEMP include a large variety of products with different technological treatments, in this review we will present data of its frequency in cooked, cured, fermented and smoked RTEMP. Preventive measures will be highlighted and discussed, namely those associated with the pathogen´s establishment and transmission from processing equipments to foods, such as accurate validation of hygiene procedures, application of "seek and destroy" method, equipment and plant design improvement, rigorous temperature control and staff training towards hygiene. Finally, we will also consider different technological strategies for *L. monocytogenes* control including: RTEMP reformulation with growth inhibitors use, technological process redesign with the adoption of emerging technologies, use of biocides in the food processing environment and antimicrobial packaging. *L. monocytogenes* control in RTEMP requires coordinated efforts along the food chain and should be regarded as part of a more vast process, in which public health policies play an important role in reducing the burden of listeriosis.

Keywords: *Listeria monocytogenes*, ready-to-eat meat-based foods, processing environment, retail, proactive measures, control, emergent technologies

INTRODUCTION

Listeria monocytogenes is a Gram-positive non-spore forming rod, facultatively anaerobic, extensively dispersed in the environment, that has been isolated from a variety of sources, including soil, vegetation, silage, faecal material, sewage and water (Alvarez-Ordóñez et al. 2015). Although the bacterial genus *Listeria* currently encompasses several species, human listeriosis is almost exclusively caused by the species *L. monocytogenes* and, exceptionally, by *L. ivanovii,* which is infrequently incriminated in human pathology (Tourdjman et al. 2014).

Listeriosis is a relatively rare but serious illness in humans, with high morbidity, hospitalization and mortality in vulnerable populations. Pregnant women, the unborn, newborns, the elderly and immune-compromised people are most commonly affected, with high mortality rates associated. Symptoms range from mild influenza-like or gastrointestinal illness to miscarriage, stillbirth, septicaemia, meningitis or encephalitis (Gillespie et al. 2010). Listeriosis is almost entirely transmitted through the ingestion of contaminated foods, predominantly ready-to-eat foods (RTE) (Lamont et al. 2011).

Unlike the majority of food-borne pathogens, *L. monocytogenes* is able to grow in low moisture content and high salt concentration foods, and most importantly, it can grow at refrigeration temperatures (Noordhout et al. 2014). Additionally, *L. monocytogenes* readily produces biofilms, thriving for long periods in food production plants (Kushawaha & Muriana, 2009).

RTE meat-based food products (RTEMP) are one of the most consumed food products around the world (Pérez-Rodriguez et al. 2010). Their long shelf-lives in refrigerated storage, as well as the fact that they do not need to undergo any heat-treatment prior to consumption, makes them an important source of food-borne disease, especially by *L. monocytogenes* (EFSA 2015).

In food processing facilities, the major source of *L. monocytogenes* contamination seems to be due to environmental events, particularly at post-processing stages (Malley et al. 2015). Processing machines, as slicers, cutters and shredders, may constitute harborage sites for *L.*

monocytogenes, acting as a contamination reservoir (Gormley et al. 2010; Henriques et al. 2014).

If proper controls are not in place, RTEMP may be contaminated at low levels during the manufacturing process, as cross-contamination during handling or slicing can occur, and *L. monocytogenes* will have the opportunity to multiply throughout products' refrigerated storage on distribution, sale and at consumer´s household (Swaminanthan and Gerner-Smidt 2007). Thus, *L. monocytogenes* is a major concern for RTEMP with an extended shelf-life. This means that the manufacturer, and to a lesser extent the retailer, have to prevent contamination and reduce growth opportunities (Todd & Notemans, 2011) to assure that *L. monocytogenes* will not grow to unsafe levels until the end of RTEMP shelf-life.

RTEMP have been the focus of several risk assessments and are specifically targeted regarding *Listeria* control by food regulatory agencies and food processors. A European survey on *L. monocytogenes* in ready-to-eat foods revealed that two percent of RTEMP were contaminated at the end of shelf-life (EFSA, 2013).

L. monocytogenes outbreaks are not as common as those caused by other food-borne pathogens, such as *Salmonella,* viruses and *Campylobacter* (EFSA 2015). Nevertheless, they usually have some seriously affected cases and even deaths, are often linked to food industries' workers failures, non-compliance with existing regulations, and have major economic consequences, especially if the products affect international trade (Todd and Notermans 2011).

This review focuses on *L. monocytogenes*' incidence in RTEMP along the food chain and comments on some of the potential prevention and control measures that can be used in the food industry to reduce/ eliminate this important pathogenic bacterium from food. Considering RTEMP's diversity and distinct technological treatments, in this review we will present data of its frequency in cooked, cured, fermented and smoked RTEMP. Preventive measures and pre-requisites will be highlighted and discussed, especially those closely associated with the pathogen´s establishment and transmission from processing equipments to foods, such as hygiene procedures validation, "seek and destroy" method application, equipment and premises design improvement, temperature control and staff training. Finally, we will also consider different strategies to control *L. monocytogenes* proposing technological process redesign by emerging technologies adoption, growth inhibitors use, biocides application in the food processing environment and antimicrobial packaging.

Coordinated efforts along the food chain are needed to control *L. monocytogenes* in RTEMP and public health policies are of upmost importance to reduce listeriosis burden.

1. Main Risk Factors for *Listeria monocytogenes* Human Infection

It is believed that listeriosis' incidence varies between two to four cases of listeriosis per million persons in most developed countries (Malley et al. 2015). Listeriosis at-risk groups include the elderly, immune-compromised people, pregnant women, unborn babies and neonates (Noordhout et al. 2014).

Current listeriosis incidence in different countries may be affected by demographic changes. In general, global population is ageing and, especially in developed countries, life expectancy is higher (Lahou et al. 2015). Presumably, these individuals will have, in their lifetime, a greater chance of developing debilitating chronic conditions, and might be regarded as at-risk group for food-borne listeriosis (Swaminanthan and Gerner-Smidt 2007).

Most *L. monocytogenes* infections have been reported in high-income countries, where incidence is low but fatality rate is high (Noordhout et al. 2014). In fact, a changing pattern of human listeriosis can be observed in Europe and in the United States of America (USA), where listeriosis is now affecting people over 65 years of age more frequently than pregnant women, and the majority of these elderly persons suffer from underlying disease(s) (Lahou et al. 2015).

Additionally, contemporary lifestyle pace influences food consumption patterns and one of the major trends is the growing preference for convenience, to which RTE are well associated (Martins and Germano, 2011).

A study performed by Gillespie et al. (2010) in the United Kingdom to relate socio-economic patterns with food-borne listeriosis, revealed that incidence was highest in those over 60 years of age, economically deprived, that ate food from local services and that lived in the most deprived areas of the country.

The majority of *L. monocytogenes* infections occurs without a clear connection to an outbreak and considered to be "sporadic" (Varma et al. 2007). In addition, listeriosis is likely to be an underreported infection due to its non-notifiable disease status in many countries and because of the absence of adequate surveillance programs (Di Pinto et al. 2010).

Even in those countries with appropriate surveillance systems, only invasive *L. monocytogenes* infections, that require hospitalization, are reported. The milder non-invasive form of listeriosis, also called febrile gastroenteritis, seems to be underreported (Todd and Notermans 2011).

A statistically significant increasing trend of listeriosis was registered in the European Union for the period 2009-2013 (EFSA, 2015). The 2015 European Union summary report on trends and sources of zoonoses, zoonotic agents and food-borne outbreaks published by the European Food Safety Authority (EFSA) and the European Centre for Disease Prevention and Control accounts 1,763 confirmed cases of listeriosis, with a notification rate of 0.44 cases per 100,000 population, representing a 8.6% increase compared with the previous year.

The most common foods involved in outbreaks belonged to the category "crustaceans, shellfish, molluscs and products thereof," followed by "cheese," "meat and meat products," "pig meat and products thereof," "vegetables, juices and products thereof." All the outbreaks involved two to four cases each, except for one outbreak related to "meat and meat products" with 34 cases. The vast majority of cases were reported to be domestically acquired and a seasonal pattern was observed with large summer peaks and smaller winter peaks (EFSA 2015).

USA's Centers for Disease Control and Prevention report on *Listeria* illnesses, deaths, and outbreaks (2013) acknowledges 1,651 cases of listeriosis with 21% of case-fatality rate. The majority of cases (58%) occurred among adults aged 65 years or more and only 14% were pregnancy-associated.

The most common foods associated with outbreaks of listeriosis were soft cheeses made from pasteurized milk, followed by raw produce. In a USA risk assessment study, deli meats were pointed as the food type posing the greatest risk for listeriosis, estimating approximately 1,600 illnesses per year attributed to RTEMP. In that study, considering listeriosis cases and deaths attributed to RTEMP consumption, approximately 83% were associated with sliced at retail RTEMP, while 14% were related to industrial pre-packaged RTEMP (Todd and Notermans 2011).

Although listeriosis may be caused by all 13 serotypes of *L. monocytogenes* (Yu and Jinag, 2014), more than 95% of human listeriosis is caused by serotypes 1/2a, 1/2b, and 4b, with serotype 4b causing by far the most cases of human listeriosis (Swaminathan & Gerner-Smidt 2007). Yet, serotype 1/2a seems to be more commonly isolated from food (Liu, 2008).

Because *L. monocytogenes* has a virulence gradient within its strains, differing in pathogenic potential and host specificity (Ward et al. 2004), more research is needed, namely on the phylogenetic and subtyping areas, using phenotypic and genotypic methods, to provide new insights on its epidemiology.

Based on subtyping, *L. monocytogenes* strains can be categorized into, at least, four genetic groups or evolutionary lineages (I, II, III and IV) with different but overlapping ecological niches (Orsi et al. 2011; Eskhan and Abu-Lail 2013). Lineage I and II include the majority of *L. monocytogenes* isolates (Gray et al. 2004). Lineage I strains are clonal, include serotypes 1/2b, 3b, 3c, and 4b and seem to be overrepresented among human clinical cases, whereas lineage II strains are more diversified due to horizontal gene transfer, include serotypes 1/2a, 1/2c, and 3a, are commonly isolated from foods and food-related environments and are widespread in the natural environment, being frequently isolated from animals with listeriosis and sporadic human cases (Gray et al. 2004; Eskhan & Abu-Lail, 2013). Lineage III and IV strains are rare and predominantly isolated from food-production animals and include serotypes 4a, 4b and 4c (Orsi et al. 2011), being underrepresented among human clinical cases and foods (Eskhan & Abu-Lail, 2013).

The preponderance of lineage II strains in foods and food-related environments might be explained by an increased capacity to survive and persist in food premises and equipments (Orsi et al. 2011). In 2000, a 1/2a *L. monocytogenes* (lineage II) strain was reported to be involved in a multi-state outbreak in USA , associated with sliced processed turkey meat and the trace-back study identified a single plant as the source of infection (Olsen et al. 2005). This strain seems to have persisted for at least 12 years in the plant premises, as it had already been identified in a sporadic human listeriosis case in 1988 (Orsi et al. 2011).

It has also been hypothesized that the difference in incidence of lineage I and II strains in foods and food-related environments is due to a different sensibility to acriflavine and to bacteriocins that might be present in enrichment mediums and food samples, giving lineage II a selective advantage (Orsi et al. 2011).

Certain *L. monocytogenes* strains possess specific mutations leading to a premature stop codon in the *InlA* gene (PMSC*inlA*), which attenuates its virulence in mammalian cells, and those strains may represent more than 30% of food isolates (Nightingale et al. 2008). That mutation in PMSC*inlA* appears to be more common in *L. monocytogenes* isolates from RTE than from human listeriosis cases, but when evaluating the distribution of PMSC*inlA* mutations among RTE product categories, Van Stelten et al. (2010) reported that the proportion of *L. monocytogenes* with or without PMSC*inlA* mutations was similar among isolates from RTE categories, with the exception of deli meats, which included a higher proportion of virulence-attenuated isolates. These last authors propose that these PMSC*inlA* mutations could be used to predict potential human health risk associated with consumption of *L. monocytogenes* contaminated foods.

2. *LISTERIA MONOCYTOGENES* INCIDENCE IN READY-TO-EAT MEAT BASED FOOD PRODUCTS: ROOT-CAUSE ANALYSIS

L. monocytogenes has been found in raw or processed food samples including dairy products, meat, vegetables and seafood (Churchill et al. 2006). *L. monocytogenes* presence in foods is influenced by several factors, but in processed and RTE, recontamination seems to be the major cause (Lambertz et al. 2012). It is important to analyze the root-cause of *L. monocytogenes* higher frequencies in RTEMP to allow for a better knowledge and management of risk factors. Several countries have a planned surveillance system to acknowledge the real incidence but, even so, fails in information occur due to lack of methods harmonization (sample size, frequency, methods of analysis) and insufficient implementation of data collecting systems, hampering data analysis and knowledge of the real *Listeria* burden (Luber et al. 2011).

Although far from ideal, several countries have implemented a *L. monocytogenes* surveillance system, integrating data from human clinical cases and food chain data, and consider two different stages: producing industry and retail establishments.

In Europe, RTEMP samples analysis by official authorities in 2013 revealed that *L. monocytogenes* was most commonly detected in RTEMP made with pork meat (3.4% of 36,511 samples), followed by those of bovine meat (2.3% of 2,575 samples), RTEMP made from broiler meat (1.6% of 5,275 samples), and by those of turkey meat (0.4% of 1,705 samples). In RTEMP made from broiler meat, 1.0% of the analyzed 2,479 samples were above the criterion of 100 colony-forming units (CFU)/g (EFSA 2015).

When considering *L. monocytogenes* frequency at the European producing industry level, 1.7% in single RTEMP samples and 2.8% in RTEMP batches were found to be contaminated. In the case of fermented sausages, all tested products were found to have less than 100 CFU/g (EFSA, 2015). Although the low frequency, similar to previous years, *L. monocytogenes* has the ability to multiply at refrigeration temperatures and RTEMP are associated with long shelf-lives, during which the pathogen may grow to unsafe levels.

In USA, *L. monocytogenes* incidence in RTEMP products at industrial level has diminished from 4.61% in 1990 to 0.32% in 2010, due to the enforcing of a zero-tolerance policy for *L. monocytogenes* in RTEMP (Malley et al. 2015).

At retail level, in Europe *L. monocytogenes* was present in 0.2% of single samples and 0.1% of batches of RTEMP, other than fermented sausages (EFSA, 2015). In the case of fermented sausages, all tested products were found to have less than 100 CFU/g (EFSA 2015). These low frequencies at retail level are similar to the ones reported in previous years and also lower than the ones obtained at the industrial level. In a European Union survey on *Listeria monocytogenes* prevalence in RTE at the end of shelf-life (EFSA 2013), 3470 samples were collected from retail establishments and the pathogen's prevalence was 2.07%. RTEMP were the third non-compliant RTE category, following fishery products and cheeses.

A risk assessment process conducted by USA's Food Safety and Inspection Service (FSIS) determined that RTEMP, particularly those sold at retail, are the food vehicle with the highest risk of being contaminated by *L. monocytogenes* (USDA-FSIS, 2010), although prevalence rates among different types of RTEMP were not statistically different. However, in a USA study, the highest attribution percentages for *L. monocytogenes* were RTE fully cooked meat (30%), RTE fully cooked poultry (25%), and RTE acidified/fermented meat without cooking (6%), being the ones sliced at retail establishments the most hazardous. Also, RTEMP with a growth inhibitor were associated with fewer illnesses and deaths (Todd and Notermans 2011).

Table 1 presents several studies conducted in different world locations to elucidate *L. monocytogenes* frequency in different types of RTEMP at industrial and retail level and with a root-cause analysis. Prevalence of *L. monocytogenes* in those food products varied between 0.3% and 38% and serotype 1/2a was by far the most reported one. When comparing *L. monocytogenes* prevalence in industrially sliced and packaged RTEMP versus retail sliced and packaged RTEMP, the retail samples were the most prevalent and in some cases, a statistically significant difference was found (Garrido et al. 2009; USDA-FSIS, 2010). Interestingly, seasonality is also referred: *L. monocytogenes* seems to be more prevalent in the warmer months of the year (Fallah et al. 2012; Modzelewska-Kapitula & Maj-Sobotka, 2014). Post-listericidal treatment contamination seems to be the major root-cause associated with processing equipment and utensils in operations such as slicing, cutting and weighing. RTEMP sliced at retail, especially the ones that do not have inhibitory organic acids or high lactic acid bacteria counts, are the products most prone to *L. monocytogenes* contamination (Luber et al. 2011). Some authors also refer temperature fails, whether in refrigerated storage (Garrido et al. 2009) or in the cooking process (Osaili et al. 2011), no use of growth inhibitors (USDA-FSIS, 2010) and poor segregation of the raw and cooked areas and personnel (Osaili et al. 2011; Henriques et al. 2014) in processing plants and retail establishments.

Since the major cause of RTEMP contamination by *L. monocytogenes* is re-contamination following a listericidal treatment from food contact surfaces and equipments that act as a potential contamination source, environmental testing is crucial (Luning et al. 2011). Thevenot et al. (2005) reported the presence of *L. monocytogenes* in dried sausages processing plants in France at all stages of the processing, before and after cleaning and disinfection. The most frequent serotype was 1/2a.

Those results were attributed to the complexity of the processing line and equipment and to the presence of organic residues on the equipments surfaces. Blatter et al. (2010) found *L. monocytogenes* serotype 1/2a repeatedly for more than 9 months in slicers, conveyor belts, tables, spattles and air blow-guns of a sandwich-producing industry in Switzerland, and reported that after cleaning and sanitizing methods' revision *L. monocytogenes* was no longer found on equipments or in the finished products. Prencipe et al. (2012) also reported *L. monocytogenes* presence in all stages of Parma ham processing units, although in different frequencies and concentrations. The same pulsotype was re-isolated in fresh hams processed in the same premise after 15 months.

The bacterial capability of persisting in the food environment has been described by Carpentier and Cerf (2011) as the ability to survive, grow and resist removal; it is commonly designated by biofilm-forming ability and largely associated with *L. monocytogenes*, not only in the food-producing industry but also in retail establishments. For this reason and owing to the ubiquitous nature of *Listeria*, it is not reasonable to expect food processing establishments to be continuously free of *Listeria* species and its presence, even at low levels, may be found associated with raw materials and ingredients (Luber et al. 2011). As abovementioned, *Listeria* are likely to be reintroduced into food processing environments and despite best efforts, it will eventually lead to contamination of RTE (Swaminanthan and Gerner-Smidt 2007).

Table 1. *L. monocytogenes* frequency in different RTEMP and root cause analysis

Year/ period	Location	Food chain stage	RTEMP type	Method	Analyzed samples (total)	Positive samples n (%)	Counts (cfu/g)	Other pertinent information	Root-cause analysis	Reference
2003-2005	Navarra, Spain	Retail and supermarkets	Patê and deli meat products (ham, chicken and turkey)	ISO 11290-1 and 2	540	24 (4.4%)	$<10^2$ cfu/g in vacuum-packed; $>10^2$ cfu/g in opened products	Higher incidence in in-store-packaged products (8.5%) than in manufacturer packaged products (2.7%); most frequently contaminated product: chicken luncheon meat (20%)	Additional handling, fails in hygiene procedures and storage temperatures (above 4°C) at retail	Garrido et al. 2009
2004	Greece	Supermarkets	Precut deli meats (slices, cubes)	Enzymatic qualitative immunoassay with biochemical confirmation	209	17 (8.1%)	$<10^1$ cfu/g	Most frequently contaminated samples: cooked ham and bacon cubes	High level of handling associated with increased risk of contamination	Angelidis and Koutsoumanis 2006
2005-2007	Belgium	Industrial processing plants	Cooked meat (pork, poultry, beef) and patês	Detection in 25 g of sample (VIDAS LMO method enzyme linked fluorescent assay) and enumeration by ISO 11290-2)	639	7 (1.1%)	$<10^2$ cfu/g	Most contaminated samples: luncheon ham	Occasional post-cooking process contamination	Uyttendaele et al. 2009
2005-2009	Sweden	Retail	Heat treated pork, turkey and beef	ISO 11290–1 and 2	507	6 (1.2%)	$<10^2$ cfu/g	Most contaminated samples: turkey meat followed by roast beef; most common serotype: 1/2a	Contamination during processing and handling	Lambertz et al. 2012
2006	United States of America	Retail	Cured and uncured poultry, pork and beef	Detection in 25 g of sample (FSIS standard laboratory method) and quantification [FSIS protocol 9-tube Most Probable Number (MPN) method]	7040	55 (0.78%)	Prepackaged in industry ≤0.3 MPN/g; retail sliced and packaged: <0.3 to ≥110 MPN/gram.	Positive samples: 11% prepackaged in industry; 89% retail-sliced and packaged	Not available, because it was a risk assessment study, but lack of growth inhibitors is pinpointed; conclusions state that retail-sliced products are associated with 83% of all *L. monocytogenes* deaths from RTEMP	USDA-FSIS, 2010

Year/ period	Location	Food chain stage	RTEMP type	Method	Analyzed samples (total)	Positive samples n (%)	Counts (cfu/g)	Other pertinent information	Root-cause analysis	Reference
2006-2007	São Paulo, Brazil	Retail	Salami and cooked ham	ISO 11290-1 and 11290-2	260	9 (3.5%)	$2x10^1$ – $1.9x10^3$ cfu/g	Serotype 4b was the most common	Post-heat treatment recontamination in slicing and packaging	Martins and Germano 2011
2007-2009	Italy	Supermarkets	Sliced salami	ISO 11290-1; quantification by MPN/g	112	23 (20.5%)	$<10^2$ cfu/g	Sliced and vacuum-packaged at industrial level	Probable cross-contamination during slicing operation at producing industry	Di *Pinto et al.* 2010
2007-2009	Henan, China	Supermarkets and open-air markets	Cooked	Detection in 25 g of sample (National Standards of the People's Republic of China)	176	13 (7.4%)	Not determined	Most common serotypes: 1/2a and 4b	Not analyzed	Yu & Jiang, 2014
2008-2009	United Kingdom	Markets and specialty food shops	Continental, cured, fermented and dried sausages	Detection in 25 g of sample and counting (Health Protection Agency Standard Microbiological Methods; serotyping and amplified fragment length polymorphism (AFLP)	2359	6 (0.3%)	$2.2x10^2$ -$1.5x10^6$ cfu/g	Most common serotype: 1/2c	*L. monocytogenes* contamination may increase with handling and/or cutting of RTEMP prior to packaging	Gormley et al. 2010
2008-2009	Santiago, Chile	Industrial processing plants and retail	Cooked sausages and pâté	Detection in 25 g of sample (VIDAS *Listeria* DUOkit, Biomérieux)	103	39 (38%)	not determined	Most common serotype:1/2a	Not analyzed	Montero et al. 2015
2008-2009	United Kingdom (Wales)	Point of sale or service	Cooked, fermented, smoked and patê	Detection in 25 g of sample; confirmation (API *Listeria* system, Biomérieux)	2191	36 (1.64%)	$<6x10^1$ cfu/g	Most common serotype:1/2a	Not analyzed	Meldrum et al. 2010
2008-2010	Estonia	Industrial processing plants	Cooked, smoked and fried	ISO 11290-1 and 2	7217	139 (1.9%)	$<10^2$ cfu/g	Most common serotype:1/2a	Cross-contamination from processing facilities	Kramarenko et al. 2013

Table 1. (Continued)

Year/ period	Location	Food chain stage	RTEMP type	Method	Analyzed samples (total)	Positive samples n (%)	Counts (cfu/g)	Other pertinent information	Root-cause analysis	Reference
2008-2011	Jordan	Retail stores and restaurants	Mortadella, chicken-burguer, chicken-sausage, chicken-shawirma	ISO 11290-1	120	36 (30%)	Not determined	Higher prevalence in processed RTEMP than in raw meats (data not shown)	No segregation of raw and cooked areas and personnel; fails in the designed time/temperature cooking step; cross-contamination during processing	Osaili et al. 2011
2009-2011	Poland	Industrial processing plants	Cooked and smoked pork sausages	ISO 11290-1 and 2	1068	19 (1.8%)	<10^2 cfu/g	Seasonality of *L. monocytogenes* presence: more common in spring and summer months	Not analyzed	Modzelewska-Kapitula and Maj-Sobotka 2014
2010-2011	Iran	Supermarkets, retail outlets and restaurants	Poultry salads, sausages and burguers	Detection in 25 g of sample: United States Department of Agriculture selective enrichment and isolation protocol; PCR targeting *prs* and *hly* genes	88	10 (11.4%)	Not determined	Most common serotypes: 4b and 1/2a; 4b serotype was most common during cold season while serotype 1/2a was predominant in warm season	Cross-contamination during processing and handling	Fallah et al. 2012
2011-2012	China (South)	Retail	Cooked	Detection in 25 g of sample (National Standards of the People's Republic of China, with minor modifications); PCR targeting *hly* gene; quantification by MPN/g	119	6 (5%)	<10 MPN/g;	Most common serovars: II.1 and II.2	Cross-contamination during additional handling at the retail level (open-air slicing, weighing, and packaging)	Chen et al. 2014
2012	Parma, Italy	Industrial processing plants	Dry-cured ham	ISO 11290-1 and 2	708	14 (2.0%)	<10^2 cfu/g	Most frequent serotype: 1/2c; no evidence supporting cross-contamination from raw materials (pork carcasses)	Cross-contamination from processing surfaces and instruments; occurrence of persistent strains over time	Prencipe et al. 2012

Year/ period	Location	Food chain stage	RTEMP type	Method	Analyzed samples (total)	Positive samples n (%)	Counts (cfu/g)	Other pertinent information	Root-cause analysis	Reference
2013	Nanjing, China	Supermarkets and open-air markets	Sauce-pickled, cured, smoked, roasted, dried, deep fried and prepared	Detection in 25 g of sample (China national food safety standard methods - Food microbiological examination); PCR targeting *hly*A gene	628	33 (5.3%)	Not determined	Most frequent serotypes: 1/2a, 3a and 1/2b, 3b; prevalence was higher in open-air markets than in supermarkets	Cross-contamination during processing and handling (slicing, weighing and packaging)	Wang et al. 2015
2014	Sardinia, Italy	Industrial processing plants	Fermented sausage	ISO 11290-1 and 2	50	4 (8%)	$<10^2$ cfu/g	Most common serotype:1/2c	Food contact surfaces are important harborage sites	Meloni et al. 2014
2014	Portugal	Industrial processing plants	Cooked, fermented, cured and smoked RTEMP of pork, beef and poultry;	ISO 11290-1 and 2	20	5 (20%)	$<10^2$ cfu/g	Most common serogroups: IIb and IVb	Incorrect food surfaces cleaning and disinfection, no preventive maintenance of food contact equipments, misuse of personal protective equipment	Henriques et al. 2014

RTEMP handling prior to packaging may increase *L. monocytogenes* contamination risk (Lambertz et al. 2012). Notably, slicing machines and cutting utensils are recognized as important contamination sources of RTEMP at the sale points (Pérez-Rodríguez et al. 2010). Lower microbiological quality of RTEMP at retail level has been related with small establishments, such as delicatessen and butchers. This seems to be due to improper food handling practices, insufficient food safety training and awareness, but also with financial constraints (Pérez-Rodriguez et al. 2010).

In a United Kingdom wide study, Gormley et al. (2010) reported that RTEMP with more than 100 CFU/g of *L. monocytogenes* were significantly associated with pre-packed meats sliced at retail, although these specialty meats were all stored below 8ºC. These authors sustain that *L. monocytogenes* levels below 100 CFU/g could also be hazardous if the product is able to support the pathogen's growth, and the remaining shelf-life is long enough with inconsistent storage temperatures at consumer's household.

3. Prevention and Control of *Listeria monocytogenes*

3.1. Proactive Methods for *L. monocytogenes* Control

Nowadays, food safety assurance is based on proactive methods to control pathogens such as *L. monocytogenes*. These proactive methods include prerequisites gathered in the so-called Codes of Good Hygiene Practices, preconized by sectorial food chain stakeholders, which are the base of success for the implementation of the preventive methodology known as hazard analysis and critical control points (HACCP). The prerequisites include premises layout and equipment design, personnel hygiene and staff training, rigorous temperature control, suppliers' selection and control, hygiene program, preventive maintenance, pest control, water supply and waste management, and their planning and implementation are crucial (CAC, 2007; Fraqueza & Barreto, 2015).

It is known that the processing environment, food handlers, incoming raw materials or even processed ingredients and products are frequent sources of *L. monocytogenes* contamination of RTEMP (Carpentier and Cerf 2011; Gomes-Neves et al. 2007; Lahou et al. 2015; Malley et al. 2015). Belessi et al. (2010) described in their studies that *L. monocytogenes* has been found to survive in plant materials for up to 10–12 years, which could be a potential resistance mechanism to antimicrobial agents, biocides and heat.

Henriques et al. (2014) assessed RTEMP producing industries in Portugal regarding their food safety management system with food and environment samples collection. All the assessed industries revealed a good or satisfactory level of prerequisites and HACCP-based procedures implementation, but *L. monocytogenes* frequency was high and related with those industries presenting higher audit scores. This picture seems contradictory, but particular reasons may explain *L. monocytogenes* presence: protective clothing used outside the production site; misuse of personal protective equipment, which suggests poor effectiveness of staff training program, incorrect food surfaces hygienization method, absence of sanitizers rotation, no isolation/ zoning of critical operations, no preventative maintenance of food contact equipments, food processing equipments design not allowing a correct hygienization, poor fail analysis and insufficient corrective measures efficacy evaluation.

Selected prerequisites will be assessed considering their major impact on *L. monocytogenes* environment persistence in RTEMP industry and retail establishments, as well as on RTEMP final products.

3.1.1. Plant Layout and Equipment Design Improvement

The design and construction of the site surrounding the factory is the first outer barrier of contamination (Wierenga & Holah, 2003). Food establishments should be located away from environmentally polluted areas, prone to pests infestations and of industrial activities that might pose a serious threat of contaminating foods (FAO/ WHO 2009).

The building structure is the second major preventative barrier in hygienic design, and the factory should be constructed with the aim of preventing contamination or deterioration of raw materials, processing facilities and final products (Melero et al. 2015).

The final set of barriers to control contamination is that within the factory itself, where two levels must be considered: the first separates processing from non-processing areas and the second separates high-risk from low-risk processing areas. High-risk areas also called "clean rooms" are those where the food products have already been subjected to a decontamination or preservation treatment, and thus there is a risk of product contamination (Melero et al. 2015). These "clean rooms" have restricted access, where only certain workers may enter and where the air-flow has positive pressure (Holah 2003).

In RTE processing units, compartmentalization is vital to protect the product from microbiological post-processing contamination. The processing unit design must permit that the flow of food, personnel, air and waste can proceed in the right direction. As raw materials and ingredients become incorporated into food products, they should move from the lower hygiene to the higher hygiene areas. Ideally, when there is further handling in RTEMP after post-processing, it should occur in a "clean room," in order to prevent re-contamination after the lethality treatment. Poorly designed equipments as well as worn, or even new materials, have hard-to-clean sites and spots that might harbor pathogenic bacteria that later become detached from them, contaminating the product (Aarnisalo et al. 2006). Specifically, hollow parts, crevices and cracks in flooring or conveyor belts materials (Carpentier and Cerf 2011), and places out of reach of mechanical cleaning are the most common spots for microbial harboring and shedding. Preventative maintenance of these equipments should be planned and scheduled at short periods, to verify and eliminate the existence of accumulated soils. When choosing food processing equipments, one important factor to consider is the ease of disassembling to allow for efficient hygienization. The European Hygienic Engineering and Design Group (www.ehedg.org), as well as other similar international organizations, have prepared guidelines for food processing machinery design and construction. These stress the importance of choosing non-toxic and resistant construction materials, designed with rounded corners and properly sealed joints, to ensure proper cleaning and disinfection procedures. These characteristics are of primary importance to avoid *L. monocytogenes* biofilms formation, giving rise to subsequent cross-contamination events.

3.1.2. Rigorous Temperature Control

In the food industry, thermal processing is traditionally used to inactivate pathogenic and spoilage microorganisms (Franco-Vega et al. 2015). Food products that have been subjected to an adequate heat-treatment during processing are free of vegetative pathogens and, depending on the treatments, of spore-formers being generally regarded as safe (Reij & Den

Aantrekker, 2004). If no manipulation of the food product occurs until its consumption and if the cold chain is respected, the risk of being a pathogen vehicle is inexistent. Nevertheless, *L. monocytogenes* can grow at refrigeration temperatures (4–15°C), thriving along the food chain even if advanced refrigeration technology for food distribution is used. It has been demonstrated that *L. monocytogenes* growth was reduced at temperatures below 0ºC (Lee et al. 2014). Low temperatures during manufacturing and distribution are of upmost importance to prevent *L. monocytogenes* from colonizing food environment surfaces, contaminating final products and to survive and multiply during product's shelf-life. This prerequisite is related to environmental processing temperature and the raw materials and final RTEMP storage steps in their flow process. The definition of temperature critical limits is important and should be associated with a monitoring frequency, recording and verification to demonstrate conformity of the processing step, to control *L. monocytogenes* multiplication.

3.1.3. Staff Training

At retail and food service level it is common to find a high turn-over of personnel, temporary workers, and unspecialized labor. These facts compromise the development of a regular training program, also hampering the food safety performance of those establishments (Lahou et al. 2015).

Survey studies on the knowledge of biological food hazards conducted in Turkey, South Africa and the United Kingdom revealed a general lack of knowledge on refrigeration temperature ranges, cross-contamination and personal hygiene (Gomes-Neves et al. 2007).

Food hygiene training in food industry and retail establishments is a legal requirement in European member states (European Commission 2004). Seaman and Eves (2010) propose that training activities closely associated with the processing environment would be more appropriate than food hygiene courses, usually provided away from the workplace, using solely knowledge-based assessment techniques. This is even more obvious when related to the processing equipments cleaning and sanitizing procedures. Usually, food handlers are more receptive to problem-solving measures training if a hazard is detected, as *L. monocytogenes*, than when a regular cleaning and sanitizing procedure is presented as a standard operating procedure. However, as described before, knowledge is not enough when it is not associated with proper professional attitude and experience. Hygienic knowledge may be acquired in training programs, but the competence is evidenced by hygienic attitudes and practices. As abovementioned, this can be a weakness in the food safety system, contributing to persistent cross-contamination with *L. monocytogenes*, even in industries classified with high hygienic score. Maintenance personnel, although not being food handlers, should also be considered in hygiene training courses programs. These workers will touch various surfaces in contact with unpacked products. Also, in order to eliminate contamination niches and hard to reach places in particular equipments, maintenance personnel will dismantle these machines in preparation for cleaning procedures and reassemble it after cleaning in order to maintain the production flow (Aarnisalo et al. 2006).

3.1.4. Accurate Hygiene Program and Its Validation Regarding L. monocytogenes

Recontamination of otherwise sound products through contaminated surfaces has been reported by several authors and is a major issue in RTE. As shown in Table 1, re-contamination by processing surfaces is the most commonly attributed cause for RTEMP contamination, whether from unclean or inadequately cleaned surfaces and pieces of equipment.

The presence of organic and inorganic material on a surface affects its ease of cleaning and may interfere with the activity of cleaning and disinfecting agents, by physically and chemically protecting microorganisms. Also, this organic soil may also act as a potential nutrients reservoir for microorganisms, enabling multiplication and an increase on the surface contamination (Whitehead et al. 2009), leading to the formation of ecological niches and biofilms, which can act as a source of cross-contamination within food premises. So, inadequate surfaces cleaning and disinfection procedures is an important risk factor of food contamination (Gómez et al. 2012).

Carpentier and Cerf (2011) state that because harborage sites almost inevitably exist in the premises, even in new floors, these should cleaned and sanitized before the equipment. This way, undesirable bacteria that may have been dislodged from harborage sites during this first hygienization step can be targeted by the second cleaning and sanitizing step on the outer surfaces of the equipment.

In the food industry, sanitizing is commonly used as a part of the hygiene program, being frequently applied after the cleaning procedures, in which detergents are used with or without physical means to remove and dislodge organic soil, as sanitizing procedures are mainly affected by organic matter presence. To choose an adequate sanitizer for food contact surfaces, several factors must be assessed: its availability, its effectiveness against microorganisms, its penetrative power in biofilms, its stability over time, its toxicity and odors (regarding personnel), the water hardness, and the cost-effectiveness of the product (Gaulin et al. 2011). However, sanitizers are frequently chosen for their commercial availability instead of being chosen according to prevailing circumstances in the processing environment and to "in-house" microorganisms, which will be repeatedly exposed to those substances, sometimes below the optimum concentrations. This is particularly relevant for sanitizers that may persist for long periods, such as quaternary ammonium compounds and phenolic-based sanitizers, for which tolerance development over time may be promoted by repeated exposure and/or prolonged environmental persistence (Gandhi and Chikindas 2007; Allen et al. 2014).

Quaternary ammonium compounds, chlorine and iodofors, commonly used in the food industry, have been shown to be effective against *L. monocytogenes* cells in suspension (Aarnisalo et al. 2007). But when considering *L. monocytogenes* biofilms, resistance of attached bacteria to biocides has been described (Chavant et al. 2004). This seems to be associated with the mechanical protection exerted by the synthesized exopolysaccharides and the surrounding nutrients, although intrinsic physiological factors as the adaptation of biofilm cells to stresses is also hypothesized (Belessi et al. 2011). So, the susceptibility of *L. monocytogenes* cells to sanitizers also depends on environmental factors such as pH, water activity, temperature and nutrient composition (Shi and Zhu, 2009; Kushawaha and Muriana 2009). Sanitizer rotation can be used as a preventive measure to avoid selecting resistant microbial populations. The active compounds in the counter-cycle sanitizer should be different from the ones present in the routinely and day-to-day hygiene procedures, in order to cover a larger spectrum of microorganisms. The frequency of application of the second sanitizer should be considered in the hygiene program. Cleaning and sanitizing procedures can be assessed by the seek and destroy (S&D) process. This is a method for identifying, managing and verifying the effectiveness of the hygiene program that can be applied to a single machine, a specific processing area or to the whole premises. Using a systematic approach, it allows for harborage sites identification, where microorganisms might survive or persist, despite cleaning and sanitation procedures. The S&D process has been successfully applied to control *L. monocytogenes* in processing plants and results have been published in peer-reviewed articles (Malley et al. 2015).

The regular monitoring of food-processing environment for *L. monocytogenes*, or other relevant pathogens to the produced foods, is important to provide information on their incidence and their potential presence in finished products, and can be complemented by monitoring hygiene indicators such as *Enterobacteriaceae* (Henriques et al. 2014). *L. monocytogenes* environmental testing should be regarded as an important part of the food safety management system of RTE producing plants.

Processing environment sample analysis supports sanitizers selection, provides information on the efficiency of cleaning and disinfection procedures and on the efficacy of implemented preventive measures, such as premises hygienic zoning and personnel and goods hygienic flows (Reij and Den Aantrekker 2004). Tompkin (2002), proposed a *Listeria* control program for RTE processing environments with six strategies: a) prevention of the establishment and growth of *Listeria* species in niches or other sites that can lead to contamination of RTE; b) sampling program implementation to assess the control program; c) rapid and effective response when the sampling program yields positive results for *Listeria* species; d) verification by follow-up sampling to ensure that the contamination source has been identified and corrected; e) short-term assessment of the last 4-8 samplings to facilitate early detection of problems and trends; and f) long-term assessment at appropriate intervals (quarterly, annually) to identify widely scattered contamination events and to measure overall progress towards continuous improvement.

3.2. Advanced Control Technologies

While there is the need to have appropriate preventive measures at all stages of the food chain, specific control measures at particular points of RTEMP processing are needed to eliminate or control *L. monocytogenes* without severe modifications of those products sensorial characteristics. Processing of RTEMP involves lethality treatments that, if done properly, should eliminate expected hazardous infectious levels of the potential pathogens (Sofos and Geornaras 2010; Gormley et al. 2010). For that purpose, apart from traditional technologies currently applied to RTEMP based on the control of water activity, pH, oxide-reduction potential, temperature, relative humidity and gaseous composition of atmosphere, some emergent technologies can be successfully applied. Ohmic heating, microwaving, high isostatic pressure, pulsed-light, electron beam and gamma irradiation, ultrasound (high power ultrasound and thermoultrasonic treatments), biopreservation and active packaging will be addressed in this chapter and their conjugation as preconized by Leistner and Gould (1995) theory about hurdles technologies in order to get a higher reduction or control of the biological hazards. In Table 2 selected studies on emergent thermal and non thermal technologies applied to RTEMP to control *L. monocytogenes* are presented, also resuming the RTEMP type to which the technology was applied, its efficacy and other relevant comments.

Considerations for each of the abovementioned technologies are their current potential use in RTEMP industry, as well as future trends, will be assessed.

Table 2. Selected emergent technologies used in RTEMP technological processing and their efficacy in *L. monocytogenes* reduction or elimination

Treatment	RTEMP type	Method/ tested conditions	Other pertinent information	Efficacy	Comments	Reference
Thermal technologies						
Ohmic processing	Whole beef muscle (*semitendinosus*)	3.5 kW batch ohmic heater (15 A, 0–250 V, 50 Hz)	*L. innocua* 11288 inoculated in meat sample	*L. innocua* inactivation of 7.05 log10 (CFU/g) in 4 min of ohmic heating with rapid high temperature short time (HTST) treatment at 95±5 °C)	Ohmic heating treatments eliminated the test microorganism; HTST ohmic cooking protocol gave similar instrumental quality values to conventional cooking, with a 15 fold reduction in cooking time	Zell *et al.* 2010
Microwave processing	Beef frankfurters without added antimicrobial preservatives	Microwave oven (550 W, operated at 2450 MHz); product's surface temperature first increased to and then maintained at 65°, 75°or 85°C, ranging from 2 to 19 min (total heating time)	4-strain *L. monocytogenes* (H7763, H7776, H7778, and 46877) cocktail at 10^{10} CFU/ml surface- inoculated and vacuum-packaged	*L. monocytogenes* concentration decreased linearly with heating time: inactivation rates of 0.41, 0.65, and 0.94 log (CFU/pk)/min at the surface temperature of 65°, 75°, or 85°C.	Overall rate of bacterial inactivation was 30%–75% higher with microwave in-package pasteurization than with water immersion heating; "edge-heating effect" with 10% shrinkage and expansion in frankfurters	Huang & Sites, 2007
Non-thermal technologies						
High isostatic pressure	Sliced dry-fermented sausages with no added sodium salt	600 MPa, 5 min, 12°C	RTEMP inoculated with 5 x10^5 CFU/g of a 3-strain cocktail of L. monocytogenes (CTC1011, CTC1034 and CECT 4031)	No antimicrobial effect against *L. monocytogenes*	Probable protective effect of product's low water activity and lactate content	Marcos *et al.* 2013
Pulsed ultraviolet light	Sliced dry-cured pork meat: salchichón and loin	Samples flashed with 0.7, 2.1, 4.2, 8.4 and 11.9 J/cm^2	7-8 log cfu/mL of a 3-strain cocktail of *L. monocytogenes* (CECT 4032, CECT 7467, and Scott A); Slices were surface-inoculated and individually vacuum packaged in 60 mm polyamide/polyethylene plastic bags	In both products, *L. monocytogenes* inactivation increased with fluence, reducing from 1.5 to 1.8 log cfu/cm^2	Slight differences in instrumental color parameters observed in both products; no changes in the sensorial analysis immediately after treatment and after 30 days of storage in salchichón. In loin, perceived sensory changes immediately after treatment, disappearing along the storage	Ganan *et al.* 2013
	Sliced and vacuum-packaged cooked ham and bologna sausage	Samples treated with 0.7, 2.1, 4.2 and 8.4 J/cm^2	Superficially inoculated with *L. monocytogenes*	*L. monocytogenes* can be reduced by approximately 2 log cfu/cm^2 in RTE cooked ham and 1 log cfu/cm^2 in bologna using a fluence of 8.4 J/cm2	8.4 J/cm^2 treatment did not affect the sensorial quality of cooked ham and triples its shelf-life, while treatments above 2.1 J/cm^2 negatively influenced the sensorial properties of bologna	Hierro *et al.* 2011

Table 2. (Continued)

Treatment	RTEMP type	Method/ tested conditions	Other pertinent information	Efficacy	Comments	Reference
Gamma irradiation	Vacuum-packaged cooked chicken breast	25 kGy, total irradiation time: 4.6 h	Inoculation of *L. monocytogenes* (ATCC 7644) at 4.71 (log_{10} CFU/g	*L. monocytogenes* elimination (not detectable during 60 days of storage (detection limit < 1 log_{10} CFU/g)	Suitable for extending the shelf-life of cooked chicken breast up to 60 days without compromising its overall sensory acceptability and nutritional quality	Feliciano et al. 2014
Electron-beam irradiation	Sliced and vacuum-packed dry-fermented sausages:*salchichón* and *chorizo*	Employed radiation doses between 1 and 3 kGy	0.3 cm thick slices immersed for 10 s in the *L. innocua* NTC 11288 suspension to achieve initial loads of approximately 10^8 cfu/g	No irradiation treatment was necessary to meet the 10^2 cfu/g microbiological criterion for *L. monocytogenes*; but 1.3 kGy is enough to obtain a "zero-tolerance" food safety objective	*L. innocua* NTC 11288 is more radioresistant than five different strains of *L. monocytogenes;* Dry fermented sausages treated with ≤2 kGy had negligible sensorial (appearance, odour and taste) modifications	Cabeza et al. 2009
High power ultrasound (HPU) and high pressure carbon dioxide (HPCD)	Sliced dry- cured ham	Cycles of 2 min of HPCD + HPU and 2 min of HPCD alone at the following conditions: 10 W , 6, 8 and 12 MPa, at 22, 35 and 45 °C, for 0.5–30 min	RTEMP spiked with 50 µl of **L. monocytogenes** suspension, (concentration ~10^7 CFU/g)	12 MPa, 35 °C, 10 W, 5 min assured inactivation to undetectable level of **L. monocytogenes** spiked on the product's surface (initial concentration: 10^9 CFU/g)	HPU alone was not able to induce any microbial inactivation; HPCD + HPU treatment assured inactivation; shelf-life was assured for 4 weeks at 4 °C; no differences in pH, total acidity and color	Spilimbergo et al. 2014
Biopreservation	Sliced dry-cured ham	Nisin directly applied (200 arbitrary units/cm^2) and Nisin applied through active packaging: polyvinyl alcohol films with 200 arbitrary units/cm^2	*L. monocytogenes* CTC1034 inoculated (10^7 cells/g) onto the product's surface and	The physico-chemical characteristics of the products enabled the survival of *L. monocytogenes*, but it was significantly reduced by the presence of nisin	The effect of biopreservation was greater when nisin was applied directly to the product's surface comparing with the active packaging	Hereu et al. 2012
Lactic acid bacteria (LAB)	Frankfurters	Combination of three strains of LAB, La51 (*L. animalis*), M35 (*L. amylovorus*), and D3 (*P. acidilactici*) with a final concentration of 10^9 CFU/ml, with or without a cell free extract of the same LAB mix in stationary phase added to the pouched RTEMP	6-strain *L. monocytogenes* F4243, ATCC 19112, F2365, J2818, J0161 and F6900 cocktail at 10^6 CFU/ml were inoculated and vaccum-sealed		All treatments reduced the growth of *L. monocytogenes* by at least 0.5 log CFU/pkg at the end of 8 weeks of refrigerated storage	Koo et al. 2012
Antimicrobial packaging	Sliced dry-fermented sausages with no added sodium salt	Bags of polyvinyl alcohol films with 108 µm of thickness impregnated with nisin (450 arbitrary units/cm^2)	RTEMP inoculated with 5 x10^5 CFU/g of a 3-strain cocktail of *L. monocytogenes* (CTC1011, CTC1034 and CECT 4031)	Pronounced reduction of *L. monocytogenes* counts during refrigerated storage	The pathogen was inactivated in some extent by all studied treatments during the product shelf life, even in the control, probably because of product's physico-chemical characteristics	Marcos et al. 2013

3.2.1. Thermal Technologies

3.2.1.1. Ohmic Heating

Ohmic heating is an advanced thermal processing method in which an alternating electrical current is applied through the food material and the dissipation of the electric field energy results in a rapid and uniform temperature distribution within the food (Bengston et al. 2006; Cho et al. 1999). With this technology, also called Joule heating or electro-heating, heat-processing times can be significantly reduced. Existing applications of Ohmic heating to foods include blanching, evaporation, dehydration, fermentation, extraction and thawing (Duygu and Umit 2015; Bengston et al. 2006). The microbial inactivation mechanism of ohmic heating seems to be thermal, although an additional mild electroporation process has also been suggested during ohmic heating at low frequency (50 - 60 Hz) (Cho et al. 1999).

In recent years, there have been a few studies on ohmic heating application to meats with promising results. In a challenge study with an inoculated whole beef muscle sample, ohmic heating treatments eliminated the test microorganism, *L. innocua*, below the detectable level, while maintaining the product´s sensorial attributes and reducing cooking losses (Zell et al. 2010). Ohmic processing seems to be quite promising in the RTEMP industry, especially for cooked meat products, because while it assures final product safety considering *L. monocytogenes*, it also reduces cooking losses and cuts down processing time. However, RTEMP with high level of fat could reduce conductivity with less dissipation of heat and consequent lost of efficacy in *L. monocytogenes* control (Shirsat et al. 2004).

3.2.1.2. Microwave Processing

Microwave heating can be used by the delivery of thermal energy to RTEMP inactivating *L. monocytogenes*. Microwave heating results from polar materials capacity to absorb electromagnetic energies at frequencies of 915 MHz or 2450 MHz, generating heat (Huang & Sites, 2007).

When an oscillating electric field acts in water molecules, these polar molecules will try to realign in the electric field direction at a rate of million times per second, originating molecular friction that results in the material's heating. Microwave heating in food processing is currently applied in drying, pasteurization, cooking, sterilization, thawing, and baking of foods (Chandrasekaran et al. 2013). While the technique has some deleterious results in the product sensorial attributes, such as "edge-heating" effect in which food edges suffer scorching, and some "off-flavours" in drying, several advantages of microwave are underlined. Microwave cooked products retain more taste, color, and nutritional value compared to those cooked by other conventional methods. Microwave pasteurization was found to be more effective in pathogens destruction, due to significant thermal effects magnification (Chandrasekaran et al. 2013). In a study conducted by Huang and Sites (2007) in frankfurters inoculated with *L. monocytogenes*, vacuum-packaged and submitted to the microwave treatment, the rate of *L. monocytogenes* inactivation was 30–75% faster with microwave heating than with conventional pasteurization technique. But the "edge-heating" effect occur, compromising product's sensorial quality, so more work needs to be done to better process optimization, minimizing its underlying causes.

3.2.2. Non-Thermal Technologies

3.2.2.1. High Isostatic Pressure

High isostatic pressure, also known as high hydrostatic pressure (HHP) is a non-thermal technology applied to food products, that may cause only minor deleterious changes to sensory and nutritional characteristics of foods, while effectively inactivating microorganisms (Jofré et al. 2008). Since HHP is isostatic and adiabatic, pressure is exerted uniformly and with its increase the temperature varies slightly, preventing the food from being deformed or heated, which would modify its organoleptic properties.

HPP is a very promising technology for the preservation of sliced cooked and cured RTEMP, with relatively low energy consumption, in which the treatment is usually applied to vacuum-packed RTEMP in flexible packs. However, microorganisms' resistance to HHP is very variable and dependable on the meat matrix to be treated (Hugas et al. 2002). In the study developed by Marcos et al. 2013 presented in Table 2, *L. monocytogenes* was not eliminated by HHP processing (600 MPa, 5 min, 12ºC), probably because of a protective effect of the product's low water activity and lactate content. On the contrary, Garriga et al. (2002) studied sliced, skin vacuum-packed dry cured ham samples, treated with HPP at 600 MPa, 6 min. Results showed a reduction of at least two $\log_{10}$ cycles for spoilage associated microorganisms after treatment. Also, the surviving microbiota was kept at low levels during all storage, and the treatment contributed to the organoleptic characteristics preservation during shelf-life (120 days), while preventing off-flavours, sour taste and gas formation. *L. monocytogenes* was absent in 25 g of all HPP treated samples during the storage period. Also, Kruk et al. (2011) obtained inactivation of *L. monocytogenes* to undetectable levels on chicken breast fillets with 450 MPa during 5 minutes treatment. So, HHP treatments efficacy depends on the binomial pressure and time application related to the protection of food matrix, being of major importance the optimization studies regarding spoilage microbiota and pathogen reduction with sensorial characteristics maintenance. In fact, HHP (more than 400 MPa during 154 seconds or more) in RTEMP such as dry-fermented sausages, has shown to be a major determinant to control Gram-negative spoilage microbiota, as *Enterobacteriaceae* and *Pseudomonas*, but not affecting Gram-positive microorganisms such as those of the lactic-acid bacteria (LAB) group, coagulase-negative *Staphylococci* (Alfaia et al. 2016).

Due to the absence of reliable kinetics in microbial inactivation, except under extreme conditions of heat and pressure for short treatment times, careful consideration must be given when using HHP as the critical step to assure food safety (Gill, 2012). Additionally, in current HHP industrial level equipment, the isostatic pressure ranges from 100 to 600 MPa (Hugas et al. 2002; Hereu et al. 2014). This is pointed as one of the limitations of the commercial application of HHP to the control of vegetative Gram-positive bacteria.

In order to increase the shelf-life and food safety of pressurized products, HHP is being studied in combination with other technologies, such as bacteriocins or other natural antimicrobials (Rastogi et al. 2007).

3.2.2.2. Pulsed-Light

Pulsed ultraviolet light, also known as pulsed-light, is a non-thermal food processing technique that involves discharge of short flashes (10^{-3}– 10^{2} ms) of intense, broad-spectrum light (200 – 1100 nm), with high voltage electric pulses (up to 70 Kilovolt/cm) into the food product, using a Xenon lamp (Abida et al. 2014). Although the penetration depth of ultraviolet radiation is very low, it does have potential to decontaminate the surfaces of RTEMP (Gill, 2010). Microbial inactivation is mainly attributed to ultraviolet-C component photochemical damage, although photo-thermal damage has also been proposed (Takeshita et al. 2003)

Few studies have been developed so far in RTEMP, but Sommers et al. (2009) reported a reduction of 1.9 log of *L. monocytogenes* when using a treatment dose of 4 J/cm2 in frankfurters. Hierro et al. (2012) achieved a reduction in *L. monocytogenes* of approximately 1 log cfu/cm^2 with the highest pulsed-light fluences assayed (8.4 and 11.9 J/cm^2) in beef and tuna carpaccio, however, these doses compromised the sensory quality in the short-term shelf-life of the products. Other authors have reported a negative effect in the colour, texture and oxidative stability of ham slices impacted by pulsed-ultraviolet light (Gill 2012).

3.2.2.3. Electron-Beam Irradiation and Gamma Irradiation

The use of ionizing radiation for processed foods decontamination, including RTEMP, is well established. In food processing, irradiation treatment exposes food to free energy presented as gamma rays obtained from Cobalt-60 and Cesium-137 radioisotopes, machine-generated X-rays (up to 5 MeV) and high-energy accelerated electrons (8-10 MeV), being the radiation exposure measured by dose and expressed as kGy. For RTEMP irradiation treatment doses ranging from 1 to 10 kGy are usually used (Lung et al. 2015).

Gamma ray exposures ranging from 2.45 to 3.75 kGy on frankfurters, bologna, ham and deli turkey meat were able to reduce 5-log of *L. monocytogenes* (Gill, 2012). In a study of Feliciano et al. (2014), gamma irradiation was successfully applied to cooked chicken breast Adobo, a Filipino ethnic recipe, achieving *L. monocytogenes* elimination during 60 days of storage without sensorial characteristics alteration (Table 2).

Electron-beam irradiation of turkey meatloaf coupled with antimicrobial compounds, such as potassium benzoate with sodium lactate or sodium diacetate with sodium lactate revealed to be effective in *L. monocytogenes* growth inhibition (Zhu et al. 2009). In another study, Cabeza et al. (2009) calculated the survival equation for *L. innocua* NTC 11288 (more radioresistant than five strains of *L. monocytogenes*) for sliced and vacuum-packed dry-fermented *salchichón* and *chorizo* sausages, obtaining a D_{value} of 0.47 and 0.53 when applying electron-beam irradiations of 1.12 and 1.27 kGy, respectively (Table 2). In fact, in fermentative products the application of this electron-beam technology could be soft and effective because it is coupled with biopreservation in these products. The fact that no negative sensorial changes were noticed could turn the application of this technology promising, particularly to sliced fermented products.

However, this is not applicable to other RTEMP products since studies referred the development of negative sensorial changes, such as discoloration and lipid oxidation, which are mainly due to the increase of free radical reactions rate with higher irradiation dosages. To reduce these undesirable effects, the addition of antioxidants or low oxygen packaging has been proposed for irradiated RTEMP (Gill, 2012).

These non-thermal based irradiation technologies are not well regarded and accepted by public opinion and consumers, although its safety has been recognized (Roberts, 2014).

3.2.2.4. Ultrasound Technology

High power ultrasound or "high intensity ultrasound" refers to sound waves with low frequencies of 20 kHz to around 1 MHz, with a sound intensity of 10 - 1000 W/cm^2. It is used in food processing, especially in liquid foods, in operations such as emulsion generation, drying, microbes and enzymes inactivation, heat and mass transfer enhancement and biological components separation (Pingret et al. 2013). Inactivation of microorganisms is due to cavitation-generated events, such as high shear, micro-streaming, water jets, shock waves and free radicals (Zhou et al. 2012). Ultrasounds have been tested for its efficacy in poultry and fresh produce surface *L. monocytogenes* decontamination, as well as on the pathogen's biofilms removal from stainless steel (Baumann et al. 2009). Power ultrasound application in food decontamination is mainly restricted to food surfaces treatment, because ultrasound is partially transmitted into a solid medium, due to the low moisture content of solid foods. Also, microbial inactivation by ultrasound will not be effective for packaged food products (Zhou et al. 2012). Franco-Vega et al. (2015) referred a 5.5 $\log_{10}$ reduction of *L. monocytogenes* in an *in vitro* study with the application of thermoultrasonic treatment for 10 minutes with low frequency (20 kHz) ultrasound (500W, 90 µm) at 55°, 60° or 65°C. Heat inactivation is significantly enhanced by adding ultrasound during dynamic treatments.

Spilimbergo et al. (2014) applied high power ultrasound treatment coupled with high pressure carbon dioxide to sliced dry-cured ham and reported that high power ultrasound alone was not able to induce any microbial inactivation, but when combined with high pressure carbon dioxide assured *L. monocytogenes* inactivation to undetectable levels; no sensorial changes were reported (Table 2).

More studies are needed to address ultrasound treatments applied to solid foods, as well as to assess ultrasound treatments combined with other technological hurdles, in order to better understand the inactivation dynamics and the organoleptic effects of those treatments on RTEMP.

3.2.2.5. Biopreservation

Nowadays, the concept of biopreservation is becoming very popular as a powerful and natural technology to extend food shelf-life and to enhance its safety. LAB (Lactic Acid Bacteria) have major potential for use in biopreservation, because they are generally recognized as safe and constitute the main naturally occurring bacterial group in many fermented foods. The antagonistic and inhibitory LAB properties are associated with the need for nutrients, promoting competition between bacterial populations; also, LAB produce antimicrobial metabolites such as lactic and/or acetic acid, hydrogen peroxide, reuterin and bacteriocins (El-Ziney et al. 1999; Ghanbari et al. 2013). In the last years, the bioprotective potential of endogenous LAB in relation to pathogens has been highlighted. LAB have been found to be effective in inhibiting *L. monocytogenes* in cooked meat products (Hugas et al. 1998; Koo et al. 2012) and in vacuum packaged meats (Juven et al. 1998). The use of protective cultures in RTEMP industry is increasing because of the promising results obtained in *L. monocytogenes* reduction and shelf-life extension. A possible control strategy for *L. monocytogenes* is the use of bactericidal or bacteriostatic antimicrobial agents. The bacteriocinogenic metabolites produced by LAB have different concentration thresholds for inhibition or inactivation. These thresholds depend on the specific targets of the antimicrobial substance, including cell wall, cell membrane, metabolic enzymes, protein synthesis, and genetic systems (Raybaudi-Massilia and Mosqueda-Melgar 2014).

Another strategy to use these natural antimicrobials is by active packaging or coatings. These have proved to be effective against microbial growth when added packaging systems used for food (Marcos et al. 2013). Nisin, a bacteriocin produced by *Lactococcus lactis*, has shown to be effective in inhibiting the growth of a wide range of Gram-positive bacteria, including food-borne pathogens such as *L. monocytogenes* (Marcos et al. 2013) in numerous foods. Nisin permeabilizes the bacterial cell membrane, dissipating the membrane potential, leading to leakage of intracellular fluids and eventually to cell death (Tang et al. 2013).

Potassium lactate and sodium diacetate, both organic acids salts, are widely used in the RTEMP industry as antimicrobial agents since they are naturally produced by LAB and they enter in the concept of biopreservation. These organic acids cross the cell membrane and reduce intracellular pH, impacting cell metabolism and resulting in reduced growth (Tang et al. 2013).

3.2.2.6. Active Packaging

Active packaging technologies include a large variety of possibilities and are gaining relevance in the RTEMP producing industry. In active packaging, the package, the product and its environment interact to improve its food safety, sensory properties and shelf-life (Realini and Marcos 2014). Important active packaging systems include oxygen scavengers, carbon dioxide emitters/absorbers, moisture absorbers, ethylene absorbers, ethanol emitters, flavor releasing/absorbing systems, time-temperature indicators, and antimicrobial containing films (Ozdemir and Floros 2004).

Antimicrobial packaging appears to be one of the most promising applications of active food packaging technology, allowing for spoilage microorganisms and pathogens control (Chen and Brody 2013). As a result, shelf-life expectancy is significantly extended and the microbial quality and safety of active packaged food products is improved. Some active packaging solutions use multiple function active systems, such as the combination of oxygen scavengers with carbon dioxide and/or antimicrobial releasing systems, while others have slow releasing behavior (Chen and Brody 2014; Ozdemir and Floros 2004). Although oxygen scavengers and moisture absorbers are the most important commercial sub-categories of active packaging, a rapid growth is anticipated for antimicrobial packaging in the near future, especially with LAB and nisin, when cost and performance factors limitations are overcome (Realini and Marcos 2014). Marcos et al. 2013 (Table 2) was able to reduce *L. monocytogenes* growth during shelf-life of sliced dry-fermented sausages active packaged in polyvinyl alcohol films impregnated with the bacteriocin nisin. In another study using Wiener sausages packed in polymer films containing lactocins, Blanco Massani et al. (2014) were able to demonstrate growth control of *L. innocua.*

4. Public Policies Regarding *Listeria monocytogenes* Prevention

Human listeriosis notification is mandatory in most of the 28 European member states, except for three, where notification is based on a voluntary system, and one member state, where no surveillance system exists (EFSA 2015).

When considering the notification of *L. monocytogenes* in foods, it is only mandatory in 12 European member states (EFSA 2015). These discrepancies will not allow for an overall assessment of the relative risks posed by each type of RTE (Luber et al. 2011).

Due to differences in sampling and epidemiological surveillance systems, listeriosis real numbers are still a mere supposition. Until now, it has not been possible to have a realistic perception of listeriosis and of *L. monocytogenes* presence in foods and along the food chain, in order to establish priorities of prevention and control. For this reason, the enforcement of mandatory preventive policies are accomplished with the establishment of RTEMP sampling plans, harmonization of sampling and analytical method and food safety microbiological criteria limits and levels. This harmonization is necessary and helpful for RTEMP international trade.

However, recommendations and even official regulations for the acceptable levels of *L. monocytogenes* in RTE vary throughout the world. For example, in USA and Canada a "zero-tolerance" (i.e., no *L. monocytogenes* in two 25-g samples) is currently employed (USA-CDC, 2013; Health Canada, 2010). In European Union member states, food safety criteria for *L. monocytogenes* in RTE requires absence of the pathogen in 10 samples of 25 g of RTE products intended for infants and for special medical purposes. In other RTE products than those intended for infants and special medical purposes, different criteria are considered: in those RTE unable to support *L. monocytogenes* growth, levels should be below 100 CFU/g throughout product's shelf-life (n = 5); while in RTE able to support *L. monocytogenes* growth, the pathogen must not be present in 5 samples of 25 g samples at the time of leaving the production plant and during product's shelf-life the pathogen should be below 100 CFU/g (n=5) (European Commission, 2005). To assist industrial operators in conducting *L. monocytogenes* challenge testing to assess whether a food product is able to support the pathogen growth and to determine its growth throughout product's shelf-life, a guidance document was developed, complementing food microbiological criteria regulation (Alvarez-Ordóñez et al. 2015).

Conclusion

Detailed information on *L. monocytogenes* incidence in different food categories, but especially in RTE, together with the fragmentary and scarce data on the real human listeriosis burden, complicates public health policies organization regarding this important food-borne disease. The ubiquitous nature of *L. monocytogenes* in the food-chain continuum, affecting different sized industries, retail and food service establishments, with different processing realities, hygienic status and food safety performances hampers the root-cause analysis. Nevertheless, different surveys and surveillance reports refer the impact of post-processing recontamination as the major cause of *L. monocytogenes* presence in RTEMP.

Proactive measures in food production, such as good hygiene and manufacturing practices and HACCP programs implementation play an important role in preventing *L. monocytogenes* spread in the food environment. Also, traditional control measures, such as thermal processing, curing and drying commonly applied to RTEMP products have proved to be insufficient given the current consumer's lifestyle needs and demand for convenient foods, that impose these products to be cut, sliced, shredded and prepared in easy to use and consume ways. So, emergent technologies, as post-processing irradiation, pulsed-light, microwave, ohmic heating, biopreservation and active packaging treatments, play a vital role in enhancing RTEMP safety and, concomitantly, protecting consumer's health. Currently, the combination of two or more emerging technologies seems to be the most appropriate choice. Still, more research is needed to overcome limitations related to the negative effects on RTEMP sensorial attributes of most of the bactericidal levels of emerging technologies, and to allow for the commercial and industrial application of these novel food processing technologies. Also, the assessment of new post-lethality treatments application to control *L. monocytogenes* in RTEMP is needed, with the aim of minimizing the application of chemical additives.

L. monocytogenes control in RTEMP requires coordinated efforts along the food continuum involving industry, academia, regulators and authorities but also the consumer.

Acknowledgments

The authors wish to thank the Centro de Investigação Interdisciplinar em Sanidade Animal (CIISA) for logistic support and the financial support from the Fundação para a Ciência e a Tecnologia (FCT) project "Portuguese traditional meat products: strategies to improve safety and quality" (PTDC/AGR-ALI/119075/2010) and research grant SFRH/BD/70711/2010.

References

Aarnisalo, K., Tallavaara, K., Wirtanen, G., Maijala, R., & Raaska, L. (2006). The hygienic working practices of maintenance personnel and equipment hygiene in the Finnish food industry. *Food Control*, 17, 1001–1011.

Aarnisalo, K., Lundén, J., Korkealab, H., Wirtanen, G. (2007). Susceptibility of *Listeria monocytogenes* strains to disinfectants and chlorinated alkaline cleaners at cold temperatures. *LWT – Food Science and Technology*, 40 (6), 1041-1048.

Abida, J., Rayees, B., Masoodi, F. A. (2014). Pulsed light technology: a novel method for food preservation. *International Food Research Journal*, 21(3): 839-848.

Alfaia, A., Alfaia, C. M., Fernandes, M. J., Fernandes, M. H., Patarata, L., Elias, M., Henriques, M., Fraqueza, M. J. (2016). Binomial effects of high isostatic pressure and time on the nutritional, sensory and microbiological characteristics of vacuum-packaged dry-fermented sausages. Submitted to Food Research International.

Alvarez-Ordóñez, A., Leong, D., Hickey, B., Beaufort, A., Jordan, K. (2015). The challenge of challenge testing to monitor *Listeria monocytogenes* growth on ready-to-eat foods in Europe by following the European Commission 2014 Technical Guidance document. *Food Research International,* 75, 233–243.

Angelidis, A.S., Koutsoumanis, L. (2006). Prevalence and concentration of *Listeria monocytogenes* in sliced ready-to-eat meat products in the Hellenic retail market. *Journal of Food Protection*, 69, 4, 938 – 942.

Baumann, A. R., Martin, S. E., Hao, F. (2009). Removal of *Listeria monocytogenes* biofilms from stainless steel by use of ultrasound and ozone. *Journal of Food Protection*; 72, 1306–1309.

Belessi, C-E. A., Gounadaki, A. S., Psomas, A. N., Skandamis, P. N. (2010). Efficiency of different sanitation methods on *Listeria monocytogenes* biofilms formed under various environmental conditions. *International Journal of Food Microbiology*, 145, S46-S52.

Bengston, R., Birdsall, E., Feilden, S., Bhattiprolu, S., Bhale, S., Lima, M. Ohmic and induction heating. In: Hui, Y. H., editor. Handbook of Food Science, Technology and Engineering. Florida, USA: CRC Press, Taylor and Francis Group, 2006; 120.

Blanco Massani, M., Molina, V., Sanchez, M., Renaud, V., Eisenberg, P., Vignolo, G. (2014). Active polymers containing *Lactobacillus curvatus* CRL705 bacteriocins: Effectiveness assessment in Wieners. *International Journal of Food Microbiology,* 178, 7–12.

Blatter, S., Giezendanner, N., Stephan, R., & Zweifel, C. (2010). Phenotypic and molecular typing of *Listeria monocytogenes* isolated from the processing environment and products of a sandwich-producing plant. *Food Control*, 21, 1519–1523.

Cabeza, M., de la Hoz, L., Velasco, R., Cambero, M., & Ordonez, J. (2009). Safety and quality of ready-to-eat dry fermented sausages subjected to E-beam radiation. *Meat Science*, 83(2), 320-327.

Carpentier, B., Cerf, O. (2011). Review - Persistence of *Listeria monocytogenes* in food industry equipment and premises. *International Journal of Food Microbiology, 145*, 1–8.

Chandrasekaran, S., Ramanathan, S., Basak, T. (2013). Microwave food processing - A review. *Food Research International*, 52, 243–261.

Chaturongakul, S., Raengpradub, S., Wiedmann, M., & Boor, K. J. (2008). Modulation of stress and virulence in *Listeria monocytogenes*. *Trends in Microbiology,* 16(8), 388-396.

Chavant, P., Gaillard-Martinie, B., Hebraud, M. (2004). Antimicrobial effects of sanitizers against planktonic and sessile *Listeria monocytogenes* cells according to the growth phase. *FEMS Microbiology Letters*, 23, 241-248.

Chen, M., Wu, Q., Zhang, J., Yan, Z., & Wang, J. (2014). Prevalence and characterization of *Listeria monocytogenes* isolated from retail-level ready-to-eat foods in South China. *Food Control*, 38, 1-7.

Chen, J., Brody, A. L. (2013). Use of active packaging structures to control the microbial quality of a ready-to-eat meat product, *Food Control*, 30 (1), 306-310.

Cho, H. Y., Yousef, A. E., & Sastry, S. K. (1999). Kinetics of inactivation of *Bacillus subtilis* spores by continuous or intermittent ohmic and conventional heating. *Biotechnology and Bioengineering*, 62(3), 368−372.

Churchill, R.L.T., Lee, H., Hall, J. C. (2006). Detection of *Listeria monocytogenes* and the toxin listeriolysin O in food. *Journal of Microbiological Methods*, 64, 141 – 170.

Codex Alimentarius Commission (CAC). (2007). Guidelines on the application of general principles of food hygiene to the control of *Listeria monocytogenes* in ready-to-eat foods (CAC/GL 61-2007). Codex Alimentarius Commission. Available at: http://www.codexalimentarius. net/input/download/standards/10740/CXG_061e.pdf Accessed 28.06.15.

Di Pinto, A., Novello, L., Montemurro, F., Bonerba, E. & Tantillo, G. (2010). Occurrence of *Listeria monocytogenes* in ready-to-eat foods from supermarkets in Southern Italy. *New Microbiologica*, 33, 249 – 252.

Duygu, B., Ümit, G. (2015). Application of Ohmic Heating System in Meat Thawing, Procedia - *Social and Behavioral Sciences,* 195, 2822-2828.

El-Ziney, M., van den Tempel, T., Debevere, J., & Jakobsen, M. (1999). Application of reuterin produced by *Lactobacillus reuteri* 12002 for meat decontamination and preservation. *Journal of Food Protection,* 62, 257–261.

Eskhan, A. O., Abu-Lail, N. I. (2013). Cellular and molecular investigations of the adhesion and mechanics of *Listeria monocytogenes* lineages I and II environmental and epidemic strains. *Journal of Colloid and Interface Science,* 394, 554–563.

European Commission. (2004). Commission Regulation (EC) No 852/2004 of the European Parliament and of the Council of 29 April 2004 on the hygiene of foodstuffs. *Official Journal L* 139, 30.4.2004, 1-23.

European Commission. (2005). Commission Regulation (EC) No 2073/2005 of 15 November 2005 on microbiological criteria for foodstuffs. *Official Journal L* 338, 22.12.2005, 1-37.

European Food Safety Authority (EFSA). Analysis of the baseline survey on the prevalence of *Listeria monocytogenes* in certain ready-to-eat foods in the EU, 2010-2011 Part A: *Listeria monocytogenes* prevalence estimates. *EFSA Journal* 2013; 11(6):3241.

European Food Safety Authority (EFSA) & European Centre for Disease Prevention and Control (ECDC). The European Union Summary Report on Trends and Sources of Zoonoses, Zoonotic Agents and Food-borne Outbreaks in 2013. 2015. *EFSA Journal* 2015; 13(1):3991. Available from: http://www.efsa.europa.eu/en/efsajournal/pub/3991.htm

Fallah, A. A., Saei-Dehkordi, S. S., Rahnama, M., Tahmasby, H., Mahzounieh, M. (2012). Prevalence and antimicrobial resistance patterns of *Listeria* species isolated from poultry products marketed in Iran. *Food Control,* 28, 327-332.

Feliciano, C.P., De Guzman, Z.M., Tolentino, L.M.M., Cobar, M.L.C., Abrera, G.B., (2014). Radiation-treated ready-to-eat chicken breast Adobo for immunocompromised patients. *Food Chemistry*, 163, 142-146.

Food and Agriculture Organization/World Health Organization (FAO/WHO). 2009. Codex Alimentarius: Food Hygiene, 4th edition. Rome: Sales and Market.

Franco-Vega, A., Ramírez-Corona, N., Lopez-Malo, A., Palou, E. (2015). Estimation of *Listeria monocytogenes* survival during thermoultrasonic treatments in non-isothermal conditions: Effect of ultrasound on temperature and survival profiles. *Food Microbiology*, 52, 124-130.

Fraqueza, M.J., Barreto, A.S. HACCP. In: Toldra, F., Hui, H. Y., Astiasaran, I., Sebranek, J., Talon, R. Handbook of Fermented Meat and Poultry, West Sussex, United Kingdom. Wiley-Blackwell; 2015; 469-485.

Ganan, M., Hierro, E., Hospital, X. F., Barroso, E., & Fernandez, M. (2013). Use of pulsed light to increase the safety of ready-to-eat cured meat products. *Food Control*, 32(2), 512-517.

Gandhi, M., Chikindas, M.L., 2007. *Listeria*: a food-borne pathogen that knows how to survive. *International Journal of Food Microbiology*, 113, 1–15.

Garrido, V., Vitas, A.I. & Garcia-Jalón, I. (2009). Survey of *Listeria monocytogenes* in ready-to-eat products: prevalence by brands and retail establishments for exposure assessment of listeriosis in Northern Spain. *Food Control*, 20, 986 – 991.

Garriga, M., Aymerich, M. T., & Hugas, M. (2002). Effect of high pressure processing on the microbiology of skin-vacuum packaged sliced meat products: cooked pork ham, dry cured pork ham and marinated beef loin (Profit Final Project Report FIT060000200066).

Gaulin, C., Le, M-L., Shum, M., Fong, D. (2011). Disinfectants and sanitizers for use on food contact surfaces-revised. Evidence review. National collaborating centre for environmental health, Vancouver, Canada. Acessed:15.06.2015. Available at: http://www.ncceh.ca//sites/default/files/Food_Contact_Surface_Sanitizers_Aug_2011.pdf

Ghanbari M, Jami M, Domig KJ, Kneifel W. (2013). Seafood biopreservation by lactic acid bacteria – a review. *Food Science and Technology*; 54, 315-324.

Gill, A. Microbial decontamination of raw and ready-to-eat meats.In: Demirci, A., Ngadi, M. O. Microbial Decontamination in the Food Industry: Novel Methods and Applications. Cambridge, United Kingdom. Woodhead Publishing Limited; 2012: 30-59.

Gillespie, I. A., Mook, P., Little, C.L., Grant, K. A., McLauchlin, J. (2010) Human listeriosis in England, 2001-2007: association with neighbourhood deprivation. *Euro Surveillance,* 15, 7–16.

Gomes-Neves, E., Araújo, A. C., Ramos, E., Cardoso, C. S. (2007).Food handling: Comparative analysis of general knowledge and practice in three relevant groups in Portugal. *Food Control*, 18, 707–712.

Gómez, D., Ariño, A., Carramiñana, J.J., Rota, C., Yangüela, J. (2012). Sponge versus mini-roller for the surface microbiological control of *Listeria monocytogenes*, total aerobic mesophiles and *Enterobacteriaceae* in the meat industry, *Food Control*, 27(1), 242-247.

Gormley, F.J., Little, C.L., Grant, K.A., de Pinna, E., McLauchlin, J. (2010). The microbiological safety of ready-to-eat specialty meats from markets and specialty food shops: A UK wide study with a focus on Salmonella and *Listeria monocytogenes*. *Food Microbiology*, 27, 243-249.

Gray, M. J., Zadoks, R. N., Fortes, E. D., Dogan, B., Cai, S., Chen, Y., Scott, V. N., Gombas, D. E., Boor, K. J., Wiedmann, M. (2004). *Listeria monocytogenes* isolates from foods and humans form distinct but overlapping populations. *Applied and environmental microbiology*, 70 (10), 5833–5841.

Health Canada. (2010). Policy on *Listeria monocytogenes* in ready-to-eat foods. Issue date: November 3, 2010. Effective date April 1, 2011. Ottawa, Ontario: Health.

Products and Food Branch, Health Canada. Available at: http://www.hc-sc.gc.ca/fn-an/legislation/pol/policy_listeria_monocytogenes_2010-eng.php. Accessed 03.07.15.

Henriques, A.R., Telo da Gama, L., Fraqueza, M.J. (2014). Assessing Listeria monocytogenes presence in Portuguese ready-to-eat meat processing industries based on hygienic and safety audit. *Food Research International*, 63, 81–88.

Hereu, A., Bover-Cid, S., Garriga, M., Aymerich, T., 2012. High hydrostatic pressure and biopreservation of dry-cured ham to meet the Food Safety Objectives for *Listeria monocytogenes*. *International Journal of Food Microbiology*. 154, 107-112.

Hereu, A., Dalgaard, P., Garriga, M., Aymerich, T., Bover-Cid, S. (2014). Analysing and modelling the growth behaviour of *Listeria monocytogenes* on RTE cooked meat products after a high pressure treatment at 400 MPa. *International Journal of Food Microbiology*, 186, 84–94.

Hierro, E., Barroso, E., Ordóñez, J. A., & Fernández, M. (2011). Efficacy of pulsed light for shelf-life extension and inactivation of *Listeria monocytogenes* on ready-to-eat cooked meat products. *Innovative Food Science & Emerging Technologies,* 12, 275–281.

Hierro, E., Ganan, M., Barroso, E., Fernández, M. (2012). Pulsed light treatment for the inactivation of selected pathogens and the shelf-life extension of beef and tuna carpaccio.International *Journal of Food Microbiology,* 158, 42-48.

Holah, J. T., Guidelines for the hygienic design, construction and layout of food processing factories. *Guideline No.39. Campden: Campden & Chorleywood Food Research Association;* 2003.

Huang, L., & Sites, J. (2007). Automatic control of a microwave heating process for in-package pasteurization of beef frankfurters. *Journal of Food Engineering,* 80, 226–233.

Hugas, M., Pages, F., Garriga, M., & Monfort, J. (1998). Application of the bacteriocinogenic *Lactobacilus sakei* CTC494 to prevent growth of *Listeria* in fresh and cooked meat products packed with different atmospheres. *Food Microbiology*, 15, 639–650.

Hugas, M., Garriga, M., Monfort, J.M. (2002). New mild technologies in meat processing: high pressure as a model technology. *Meat Science*, 62, 359-371.

Jofré, A., Garriga,M., Aymerich,T. (2008). Inhibition of Salmonella sp. Listeria monocytogenes and Staphylococcus aureus in cooked ham by combining antimicrobials, high hydrostatic pressure and refrigeration, *Meat Science*, 78 (1–2), 53-59.

Juven, B., Barefoot, S., Pierson, M., McCaskill, L., & Smith, B. (1998). Growth and survival of *Listeria monocytogenes* in vacuum-packaged ground beef inoculated with *Lactobacillus alimentarius* FloraCarn L-2. *Journal of Food Protection*, 61, 551–556.

Koo, O-K., Eggleton, M., O'Bryan,C. A., Crandall, P.G., Ricke, S. C. (2012). Antimicrobial activity of lactic acid bacteria against *Listeria monocytogenes* on frankfurters formulated with and without lactate/diacetate, *Meat Science*, Volume 92(4), 533-537.

Kramarenko, T., Roasto, M., Meremäe, K., Kuningas, M., Põltsama, P., Elias, T. (2013). *Listeria monocytogenes* prevalence and serotype diversity in various foods. *Food Control*, 30, 24-29.

Kruk, Z. A., Yun, H., Rutley, D. L., Lee, E. J., Kim, Y. J., Jo, C. (2011). The effect of high pressure on microbial population, meat quality and sensory characteristics of chicken breast fillet, *Food Control*, 22 (1), 6-12.

Kushawaha, K., & Muriana, P. M. (2009). Adherence characteristics of *Listeria* strains isolated from three ready-to-eat meat processing plants. *Journal of Food Protection*, 72, 2125-2132.

Lahou, E., Jacxsens, L., Verbunt, E., Uyttendaele, M. (2015). Evaluation of the food safety management system in a hospital food service operation toward *Listeria monocytogenes. Food Control*, 49, 75-84.

Lambertz, S. T., Nilsson, C., Brådenmark, A., Sylvén, S., Johansson, A., Jansson, L.-M., Lindblad, M. (2012). Prevalence and level of *Listeria monocytogenes* in ready-to-eat foods in Sweden 2010. *International Journal of Food Microbiology*, 160(1), 24-31.

Lamont, R.F., Sobel, J., Mazaki-Tovi, S., Kusanovic, J. P., Vaisbuch, E., Kim, S.K., Uldbjerg, N., Romero, R. (2011). Listeriosis in human pregnancy: a systematic review. *Journal of Perinatal Medicine*, 39, 227–36.

Lee, Y. J., Jung, B. S., Yoon, H. J., Kim, K-T., Paik H. D, Lee, J. Y. (2014). Predictive model for the growth kinetics of *Listeria monocytogenes* in raw pork meat as a function of temperature. *Food Control*, 44, 16-21.

Leistner, L., Gould, G. W. 1st edition. New York, USA: Springer Kluwer Academics / Plenum Publishers, 2002; 1-195.

Liu, D., Handbook of *Listeria monocytogenes*. 1st Edition. Boca Raton, USA: CRC Press Taylor & Francis Group, LLC; 2008.

Luber, P., Crerar, S., Dufour, C., Farber, J., Datta, A., Todd, E. C. D. (2011). Controlling *Listeria monocytogenes* in ready-to-eat foods: Working towards global scientific consensus and harmonization e Recommendations for improved prevention and control. *Food Control*, 22, 1535-1549.

Lung, H-M., Cheng, Y-C., Chang, Y-H., Huang, H-W., Yang, B. B., Wang, C-Y. (2015). Microbial decontamination of food by electron beam irradiation. *Trends in Food Science & Technology*, 44 (1), 66-78.

Luning, P. A., Jacxsens, L., Rovira, J., Osés, S. M., Uyttendaele, M., & Marcelis, W. J. (2011). A concurrent diagnosis of microbiological food safety output and food safety management system performance: Cases from meat processing industries. *Food Control*, 22(3–4), 555–565.

Malley, T. J.V., Butts, J., Wiedmann, M. (2015). Seek and Destroy Process: *Listeria monocytogenes* Process Controls in the Ready-to-Eat Meat and Poultry Industry. *Journal of Food Protection*, 78 (2), 436–445.

Marcos, B., Aymerich, T., Garriga, M., Arnau, J. (2013). Active packaging containing nisin and high pressure processing as post-processing listericidal treatments for convenience fermented sausages. *Food Control,* 30, 325–330.

Martins, E. A., Germano, P. M. L. (2011). *Listeria monocytogenes* in ready-to-eat, sliced, cooked ham and salami products, marketed in the city of São Paulo, Brazil: Occurrence, quantification, and serotyping. *Food Control*, 22, 297-302.

Melero, B., Diez, A. M., Rovira, J. Basic Sanitation. In: Toldra, F., Hui, H. Y., Astiasaran, I., Sebranek, J., Talon, R. Handbook of Fermented Meat and Poultry, West Sussex, United Kingdom. Wiley-Blackwell; 2015; 443-449.

Modzelewska-Kapitula, M., Maj-Sobotka, K. (2014). The microbial safety of ready-to-eat raw and cooked sausages in Poland: *Listeria monocytogenes* and Salmonella spp. occurrence. *Food Control*, 36, 212-216.

Montero, D., Bodero, M., Riveros, G., Lapierre, L., Gaggero, A., Vidal, R. M., Vidal, M. (2015). Molecular epidemiology and genetic diversity of *Listeria monocytogenes* isolates from a wide variety of ready-to-eat foods and their relationship to clinical strains from listeriosis outbreaks in Chile. *Frontiers in Microbiology*, 6:384, 1-8.

Nightingale, K. K., Ivy, R. A., Ho, A. J., Fortes, E. D., Njaa, B. L., Peters, R. M., Wiedmann, M. (2008). inlA premature stop codons are common among *Listeria monocytogenes* isolates from foods and yield virulence-attenuated strains that confer protection against fully virulent strains. *Applied & Environmental Microbiology*, 74, 6570-6583.

Noordhout, C. M., Devleesschauwer,B., Angulo,F. J., Verbeke,G., Haagsma,J.,Kirk, M., Havelaar,A., Speybroeck, N. (2014) The global burden of listeriosis: a systematic review and meta-analysis. *Lancet Infectious Diseases*, 14, 1073–82.

Olsen, S. J., Patrick, M., Hunter, S. B., Reddy, V., Kornstein, L., MacKenzie, W. R., Lane, K., Bidol, S., Stoltman, G. A., Frye, D. M., Lee, I., Hurd, S., Jones, T. F., LaPorte, T. N., Dewitt, W., Graves, L., Wiedmann, M., Schoonmaker-Bopp, D. J., Huang, A. J, Vincent, C., Bugenhagen, A., Corby, J., Carloni, E. R., Holcomb, M. E., Woron, R. F., Zansky, S. M., Dowdle, G., Smith, F., Ahrabi-Fard, S., Ong, A. R., Tucker, N., Hynes, N. A., Mead, P. (2005). Multistate Outbreak of *Listeria monocytogenes* Infection Linked to Delicatessen Turkey Meat. *Clinical Infectious Disease*, 40 (7), 962-967.

Orsi, R. H., den Bakker, H. C., Wiedmann, M. (2011). *Listeria monocytogenes* lineages: Genomics, evolution, ecology, and phenotypic characteristics. *International Journal of Medical Microbiology*, 301, 79–96.

Osaili, T. M., Alaboudi, A. R., Nesiar, E. A. (2011). Prevalence of *Listeria* spp. and antibiotic susceptibility of *Listeria monocytogenes* isolated from raw chicken and ready-to-eat chicken products in Jordan. *Food Control*, 22, 586-590.

Ozdemir, M., Floros, J. D. (2004) Active Food Packaging Technologies, *Critical Reviews in Food Science and Nutrition*, 44(3), 185-193.

Pérez-Rodríguez, F., Castro, R., Posada-Izquierdo, G.D., Valero, A., Carrasco, E., García-Gimeno, R.M., Zurera, G. (2010). Evaluation of hygiene practices and microbiological quality of cooked meat products during slicing and handling at retail, *Meat Science*, 86: 479-485.

Pingret, D., Fabiano-Tixier, A-S., Chemat, F. (2013). Degradation during application of ultrasound in food processing: A review. *Food Control*, 31, 593-606.

Prencipe, V. A., Rizzi, V., Acciari, V., Iannetti, L., Giovannini, A., Serraino, A., Calderone, D., Rossi, A., Morelli, D., Marino, L., Migliorati, G., Caporale, V. (2012). *Listeria monocytogenes* prevalence, contamination levels and strains characterization throughout the Parma ham processing chain. *Food Control*, 25, 150-158.

Rastogi, N. K., Raghavarao, K. S. M. S., Balasubramaniam, V. M., Niranjan, K., & Knorr, D. (2007). Opportunities and challenges in high pressure processing of foods. *Critical Reviews in Food Science and Nutrition*, 47, 69–112.

Raybaudi-Massilia, R. & Mosqueda-Melgar, J. Novel Technologies for Controlling *Listeria monocytogenes* in Ready-to-Eat Products. In Hambrick, E. C., editor. *Listeria monocytogenes*: Food sources, prevalence and management strategies. New York, USA: Nova Science Publishers, Inc.; 2014; 205-229.

Realini, C. E., Marcos, B. (2014). Active and intelligent packaging systems for a modern society, *Meat Science*, 98 (3), 404-419.

Reij, M. W., Den Aantrekker, E. D., ILSI Europe Risk Analysis in Microbiology Task Force (2004). Recontamination as a source of pathogens in processed foods. *International Journal of Food Microbiology*, 91, 1-11.

Reniere, M. L., Whiteley, A.T., Hamilton, K. L., John, S. M., Lauer, P., Brennan, R.G., Portnoy, D. A. (2015). Glutathione activates virulence gene expression of an intracellular pathogen. *Nature,* 517, 170-175.

Roberts, P. B. (2014). Food irradiation is safe: Half a century of studies, *Radiation Physics and Chemistry*, 105, 78-82.

Seaman, P., Eves, A. (2010). Perceptions of hygiene training amongst food handlers, managers and training providers – A qualitative study. *Food Control,* 21, 1037–1041.

Shi, X. M., Zhu, X. N. (2009). Biofilm formation and food safety in food industries. *Trends in Food Science and Technology*, 20, 407-413.

Shirsat, N., Lyng, J.G., Brunton, N.P., McKenna, B. (2004). Ohmic processing: Electrical conductivities of pork cuts, *Meat Science*, 67(3), 507-514.

Sommers, C. H., Cooke, P. H., Fan, X., Sites, J. E. (2009). Ultraviolet light (254 nm) inactivation of Listeria monocytogenes on frankfurters that contain potassium lactate and sodium diacetate. *Journal of Food Science,* 74(3), 114-119.

Spilimbergo, S., Cappelletti, M., Ferrentino, G. (2014). High pressure carbon dioxide combined with high power ultrasound processing of dry cured ham spiked with *Listeria monocytogenes*. *Food Research International*, Volume 66, 264-273.

Sofos, J.N., Geornaras, I., 2010. Overview of current meat hygiene and safety risks and summary of recent studies on biofilms, and control of Escherichia coli O157:H7 in nonintact, and *Listeria monocytogenes* in ready-to-eat, meat products. *Meat Science*. 86, 2–14.

Swaminathan, B., Gerner-Smidt, P. (2007). The epidemiology of human listeriosis. *Microbes and Infection*, 9, 1236-1243.

Takeshita, K., Shibato, J., Sameshima, T., Fukunaga, S., Isobe, S. Arihara, K., Itoh, M. (2003) damage of yeast cells induced by pulsed light irradiation. *International Journal of Food Microbiology*, 85, 151-158.

Tang, S., Stasiewicz, M.J., Wiedmann, M., Boor, K.J., Bergholz, T.M. (2013). Efficacy of different antimicrobials on inhibition of *Listeria monocytogenes* growth in laboratory medium and on cold-smoked salmon. *International Journal of Food Microbiology*, 165, 265–275.

Thevenot, D., Delignette-Muller, M.L., Christieans, S., Vernozy-Rozand, C. (2005). Prevalence of *Listeria monocytogenes* in 13 dried sausage processing plants and their products. *International Journal of Food Microbiology,* 102, 85-94.

Todd, E. C. D., Notermans, S. (2011). Surveillance of listeriosis and its causative pathogen, *Listeria monocytogenes*. *Food Control*, 22, 1484-1490.

Tompkin, R.B. (2002). Control of *Listeria monocytogenes* in the food-processing environment, *Journal of Food Protection*, 65, 709-725.

Tourdjman, M., Laurent, E., Leclercq, A. (2014). Listériose humaine: une zoonose d'origine alimentaire. *Revue Francophone des Laboratoires*, 464, 37-44.

United States of America Centers for Disease Control and Prevention (USA-CDC). (2013). Vital sign: *Listeria* illnesses, deaths, and outbreaks – United States, 2009-2011. Morbidity and Mortality Week Report, 62 (2013), pp. 448–452. Available at: http://www.cdc.gov/mmwr/preview/ mmwrhtml/mm6222a4.htm Accessed 18.05.2015.

United States Department of Agriculture - Food Safety Inspection Service (USDA-FSIS) (2010). FSIS Comparative Risk Assessment for *Listeria monocytogenes* in Ready-to-eat Meat and Poultry Deli Meats Report. United States Department od Agriculture, Office of Public Health Science. Available at: http://www.fsis.usda.gov/wps/wcm/connect/fsis-content/internet/main/topics/science/risk-assessments/risk-assessments#RTE. Accessed 24.06.15

Uyttendaele, M., Busschaert, P., Valero, A., Geeraerd, A. H., Vermeulen, A., Jacxsens, L., Gohb, K.K., De Loy, A., Van Impe, J.F., Devlieghere, F. (2009). Prevalence and challenge tests of *Listeria monocytogenes* in Belgian produced and retailed mayonnaise-based deli-salads, cooked meat products and smoked fish between 2005 and 2007. *International Journal of Food Microbiology*, 133, 94-104.

Van Stelten, A., Simpson, J. M., Ward, T. J., Nightingale,K. K. (2010). Revelation by single-nucleotide polymorphism genotyping that mutations leading to a premature stop codon in *inlA* are common among *Listeria monocytogenes* isolates from ready-to-eat foods but not human listeriosis cases. *Applied and environmental microbiology*, 76 (9), 2783–2790.

Varma, J. K., Samuel, M. C., Marcus, R., Hoekstra, R. M., Medus, C., Segler, S., Anderson, B. J., Jones, T. F., Shiferaw, B., Haubert, N., Megginson, M., McCarthy, P. V., Graves,

L., Gilder, T. V., Angulo, F. J. (2007). *Listeria monocytogenes* infection from foods prepared in a commercial establishment: a case-control study of potential sources of sporadic illness in the United States. *Clinical Infectious Diseases*, 44, 521-528.

Wang, G., Qian, W., Zhang, X., Wang, H., Ye, K., Bai, Y., Zhou, G. (2015). Prevalence, genetic diversity and antimicrobial resistance of *Listeria monocytogenes* isolated from ready-to-eat meat products in Nanjing, China. *Food Control*, 50, 202-208.

Ward, T. J., Gorski, L., Borucki, M. K., Mandrell, R. E., Hutchins, J. Pupedis, K. (2004) Intraspecific phylogeny and lineage group identification based on the PrfA virulence gene cluster of *Listeria monocytogenes*. *Journal of Bacteriology*, 186 (15), 4994–5002.

Whitehead, K.A., Benson. P., Verran, J. (2009). Differential fluorescent staining of *Listeria monocytogenes* and a whey food soil for quantitative analysis of surface hygiene. *International Journal of Food Microbiology,* 135, 75–80.

Wierenga, G., Holah, J. T. Hygienic plant design. In: Lelieveld, H. L. M., Mostert, M. A., Holah, J., White, B. Hygiene in Food Processing; Cambridge; Woodhead; 2003.

Yu, T., Jiang, X. (2014). Prevalence and characterization of *Listeria monocytogenes* isolated from retail food in Henan, China. *Food Control,* 37, 228-231.

Zell, M., Lyng, J. G., Cronin, D. A., & Morgan, D. J. (2010). Ohmic cooking of whole beef muscle-Evaluation of the impact of a novel rapid ohmic cooking method on product quality. *Meat Science*, 86, 258-263.

Zhou, B., Lee, H., Feng, H. Microbial decontamination of food by power ultrasound. In: Demirci, A., Ngadi, M. O. Microbial Decontamination in the Food Industry: Novel Methods and Applications. Cambridge, United Kingdom. *Woodhead Publishing Limited*; 2012: 300-321.

Zhu, M.J., Mendonca, A., Ismail, H.A., Ahn, D.U. (2009). Fate of Listeria monocytogenes in ready-to-eat turkey breast rolls formulated with antimicrobials following electron-beam irradiation. *Poultry Science,* 88, 205–213.

In: Listeria monocytogenes
Editor: Tiffany Vicario
ISBN: 978-1-63483-765-1

Chapter 5

PREVALENCE OF *LISTERIA MONOCYTOGENES* IN POULTRY MARKETED IN IRAN: CHARACTERIZATION AND ANTIMICROBIAL RESISTANCE OF THE ISOLATES

***Hossein Jamali*[1], *Behrad Radmehr*[2] *and Domenico Meloni*[3]**
[1]Department of Pathology and Microbiology, Faculty of Veterinary Medicine, University of Montreal, St-Hyacinthe, Quebec, Canada
[2]Department of Food Hygiene and Quality Control, Faculty of Veterinary Medicine, Islamic Azad University, Karaj Branch, Karaj, Iran
[3]Department of Veterinary Medicine, University of Sassari, Sassari, Italy

ABSTRACT

The aim of the present study was to evaluate the prevalence of *Listeria monocytogenes (L.m)* in poultry marketed in Varamin, Tehran province, Iran. A total of 584 samples, including intestinal contents of chicken (n = 370) and turkey (n = 214) were randomly collected from different wet markets, between November 2008 to July 2010. USDA method was used for isolation of *Listeria* spp and *L. m*. The *L. m.* isolates were confirmed by API *Listeria* and PCR and were seroptyped using antisera against O and H antigens. The Kirby-Bauer disc diffusion method was used to assess the antimicrobial susceptibility to 10 antibiotics (*Tetracycline, Gentamycin, Chloramphenicol, Vancomycin, Rifampin, Pen G, Kanamycin, Erythromycin, Clindamycin and Amoxicillin-clavulanic acid)* according to CLSI. Furthermore, the virulence profile of *L.m* isolates (*inl*A, *inl*B, *inl*C, *inl*J, *act*A, *hly*A, *iap*, *plc*A, *prf*A) and the presence of antimicrobial resistance genes (*tet*A, *tet*B, *van*A, *van*B, *erm*B, *cml*A, penA) was investigated by PCR. The contamination rate with *Listeria* spp in poultry was 16%. The most frequent isolated species was *L. ivanovii* (9%). *L.m* was detected in 3% of poultry samples and serovar 1/2a (61%) was the predominant among *L.m* isolates. All virulence-associated genes were detected in 85% of the isolates. The 33% of *L.m* isolates were sensitive to all tested antibiotics: the remaining *L.m* isolates were resistant to 1 to ≥ 3 antibiotics. The presence of antimicrobial resistance genes was not completely correlated with antimicrobial resistance to *Tetracycline* and *Pen.*

The main results of the present study confirmed the high prevalence of *Listeria* spp. and the high rate of antimicrobial resistance of *L.m* isolates from poultry products marketed in Iran.

1. INTRODUCTION

The genus *Listeria* spp includes fifteen species, i.e., *L. monocytogenes, L. ivanovii, L. innocua, L. welshimeri, L. seeligeri, L. grayi, L. marthii, L. rocourtiae, L. fleichmannii, L. weihenstephanensis, L. floridensis, L. aquatic, L. cornellensis, L. riparia* and *L. grandensis* (Meloni 2014). Out of the fifteen species, *L. monocytogenes* is the major pathogen for humans. Very rare cases of infections due to *L. ivanovii* and *L. seeligeri* have been described. *L. monocytogenes* is the etiologic agent of Listeriosis, a severe disease that is characterized by two forms: febrile gastroenteritis in healthy people and life-threatening invasive infections in risk groups such as young, old, pregnant and immune-compromised (YOPI). Although the incidence of Listeriosis is low when compared with other zoonoses, it has high hospitalization rates with supplementary long-term sequelae. The highest notification rates are reported in persons aged 65 years and above and in persons aged below one year (Denny and McLauchlin, 2008). Moreover, Listeriosis is one of the leading causes of death among microbial pathogens. The majority of the infections are thought to be food-borne and associated with the consumption of contaminated meat, poultry, dairy, fish and vegetable products (Aureli et al. 2000; Gillespie et al. 2006; Public Health Agency Canada (PHAC), 2009; US Department of Agriculture's Food Safety and Inspection Service (USDA/FSIS), 2010; Todd and Notermans, 2011). Some of the morphological, physiological and epidemiological features of *L. monocytogenes*, together with the severity of human Listeriosis, may increase the public health hazard from cold-stored "ready to eat" (RTE) products (Romanova et al. 2002; Van Coillie et al. 2004; Shen et al. 2006). Poultry products are often identified as the source of foodborne pathogens (ICMSF, 1998) and raw poultry is a well recognized source of *L. monocytogenes* (Bailey et al. 1989; Uyttendaele et al. 1997). Poultry become contaminated either by the environment or from healthy carrier birds in the farm (Skovgaard and Morgen, 1988). As reported by several authors, inadequate cleaning and disinfecting procedures of poultry slaughterhouses and processing plant environments may lead to *L. monocytogenes* contamination of poultry carcasses (Uyttendaele et al. 1997; Loura et al. 2005; Fallah et al. 2012). RTE poultry products can be also contaminated after cooking by environmental cross-contamination or via workers, surfaces and equipments (Osaili et al. 2011; Fallah et al. 2012). Antimicrobial resistance (AMR), the resistance of a bacterium to an antimicrobial to which it was previously sensitive, nowadays is considered a serious risk to public health. The overuse or misuse of antimicrobials has been linked to the emergence and spread of resistant bacteria. When AMR occurs in zoonotic bacteria present in animals and food, it can compromise the effective treatment of infectious diseases in humans and in animals. Although the use of antimicrobials in poultry production has reduced the risk of infectious diseases, it may lead to dissemination of antimicrobial-resistant bacteria in the environment, such as *Salmonella* and *L. monocytogenes* (Fallah et al. 2012). Poultry products can get contaminated by *L. monocytogenes* whether the bacteria are resistant to antibiotics or not. Some resistant strains are transmitted to people through foods and may have public health consequences (Filiousis et al. 2009). This is of extreme importance because antibiotic resistant infections from foodborne pathogens can cause more severe health outcomes than infections that are not resistant to antibiotics (Krueger et al. 2014).

This study aimed to determine the prevalence of *L. monocytogenes* in poultry marketed in Varamin, Tehran province, Iran and the antimicrobial susceptibility of *L. monocytogenes* isolates to 10 different antibiotics (*Tetracycline, Gentamycin, Chloramphenicol, Vancomycin, Rifampin, Pen G, Kanamycin, Erythromycin, Clindamycin* and *Amoxicillin-clavulanic acid*). Furthermore, the serotype distribution, the virulence profile of L. monocytogenes isolates (*inl*A, *inl*B, *inl*C, *inl*J, *act*A, *hly*A, iap, *plc*A, *prf*A) and the presence of antimicrobial resistance genes (*tet*A, *tet*B, *van*A, *van*B, *erm*B, *cml*A, *pen*A) was investigated by PCR.

2. Materials and Methods

2.1. Sample Collection

A total of 584 samples, including intestinal contents of chicken (n = 370) and turkey (n = 214) were randomly collected from different wet market in Varamin, Tehran province, Iran, between November 2008 to July 2010. All the samples were identified with a univocal code and placed inside a sterile portable icechest (kept at +3°C) were transported to the laboratory and analyzed for the presence of *Listeria* spp and *L. monocytogenes* on the same day.

2.2. Isolation and Identification of Listeria Spp and L. Monocytogenes

Poultry samples were tested for the presence of *Listeria* spp using the selective enrichment and isolation protocol recommended by the United States Department of Agriculture (USDA) (McClain and Lee, 1988; Fallah et al. 2012). The putative *Listeria* spp isolates were submitted to Gram stain, catalase, and oxidase tests for confirmation of the genus *Listeria*. Biochemical characterization of all the isolates was performed using the API *Listeria* identification system (bioMerieux, Marcy l'Etoile, France). After species identification, the isolates were stored at -80°C in Brain Heart Infusion broth (BHI) (Biolife) with glycerol (15% v/v).

2.3. Confirmation of Listeria Spp and L. Monocytogenes by PCR Analysis

The phenotypic identification of *Listeria* spp and *L. monocytogenes* isolates was confirmed by a duplex-PCR based method using LI1/U1 and LM1/LM2 primers (Rossmanith, et al. 2006).

2.4. Serotyping of L. Monocytogenes Isolates

In order to identify the major *L. monocytogenes* serovars (1/2a, 1/2b, 1/2c and 4b), the isolates were serotyped by the agglutination method using commercial *Listeria* antisera (Denka Seiken Co. Ltd., Tokyo, Japan) against somatic (OI–OII, OV–OVI, OI, OIV, OVI, OVII, OVIII, OIX) and flagellar (H-A, H-AB, H-C, H-D) according to manufacturer's instructions and to the procedure of Ueda et al. (2002).

2.5. Antimicrobial Susceptibility Test

The *L. monocytogenes* isolates were tested for their susceptibility to 10 different antimicrobials: *Tetracycline* (30 µg), *Gentamycin* (10 µg), *Chloramphenicol* (30 µg), *Vancomycin* (30 µg), *Rifampin* (5 µg). *Pen G* (10 U/IE), *Kanamycin* (30 µg), *Erythromycin* (15 µg), *Clindamycin* (2 µg), and *Amoxicillin-clavulanic acid* (20/10 µg). The Kirby-Bauer disc-diffusion method using Mueller-Hinton Agar (Merck, Darmstadt, Germany) supplemented with 5% defibrinated sheep blood was used according to the Clinical and Laboratory Standards Institute (CLSI, 2006). The diameter of growth inhibition zone around each antimicrobial disc was measured after an incubation period of 24 h at + 35°C. The results were interpreted according to the CLSI (2007) recommendations.

2.6. Presence of Antimicrobial Resistance Genes by PCR Analysis

The presence of antimicrobial resistance genes (*tet*A, *tet*B, *van*A, *van*B, *erm*B, *cml*A, *pen*A) was investigated by PCR according to the protocol of Guillame et al. (2000), Lanz et al. (2003), Gebreyes and Altier (2002), and Antignac et al. (2001).

2.7. Presence of Virulence Genes by PCR Analysis

The virulence profile of *L. monocytogenes* isolates was determined by PCR through the detection of the following virulence genes (*inl*A, *inl*B, *inl*C, *inl*J, *act*A, *hly*A, *iap, plc*A, *prf*A) according to the protocol of Jamali et al. (2014) and Rawool et al. (2007).

3. Results and Discussion

3.1. Isolation and Identification of Listeria Spp and L. Monocytogenes

The prevalence and the identification of *Listeria* spp and *L. monocytogenes* in poultry marketed in Iran are shown in Table 1. The overall prevalence rate of *Listeria* spp in poultry was 16%. The prevalence in turkey and chicken samples was 16% and 15% respectively. *L. monocytogenes* was detected in 3% of poultry samples. The prevalence in turkey and chicken samples was 3.5% and 3.7% respectively. The prevalence rates of *Listeria* spp and *L.*

monocytogenes were lower than other Iranian surveys (Fallah et al. 2012). In previous studies carried out in several countries the rates of contamination with *L. monocytogenes* in poultry ranged from 12% to 38% (Ganasena et al. 1995; Gudbjörnsdóttir et al. 2004; Yücel et al. 2005; Pesavento et al. 2010; Sakaridis et al. 2011; Osaili et al. 2011).

Altogether, from all *Listeria* spp positive samples, n.94 isolates were isolated. By biochemical and molecular methods (Figure 1) the species identification was confirmed as follows: 57 were *L. ivanovii* (9.8%), 21 were *L. monocytogenes* (3.6%), 12 were *L. innocua* (2.1%), 4 were *L. seeligeri* (0.7%). *L. ivanovii* was the most common species either in chicken or turkey samples.

Table 1. Prevalence and identification of *Listeria* spp and *L. monocytogenes* in poultry samples

No. of Samples	Chicken	Turkey	Total
	370 (63.4%)	214 (36.6%)	584
Listeria spp.	61 (16.5%)	33 (15.4%)	94 (16.1%)
L. ivanovii	36 (9.7%)	21 (9.8%)	57 (9.8%)
L. monocytogenes	13 (3.5%)	8 (3.7%)	21 (3.6%)
L. seeligeri	2 (0.5%)	2 (0.9%)	4 (0.7%)
L. innocua	10 (2.7%)	2 (0.9%)	12 (2.1%)

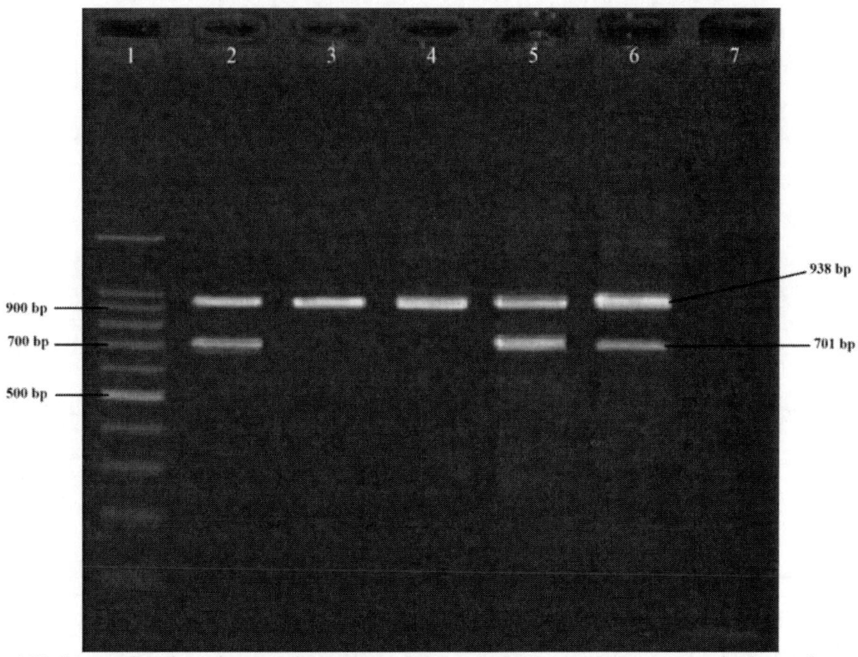

Lane 1, 100 bp molecular size marker; lanes 2,5, and 6: *L. monocytogenes* isolates; lanes 3 and 4: non-*L. monocytogenes* isolates; lane 7: negative control.

Figure 1. A reprehensive gel of PCR-amplified *Listeria* spp products. *Listeria* spp is indicated by a single band at 938 bp (LI1/U1). However, *L. monocytogenes* is indicated by two bands at 938 bp and 701 bp (LM1/LM2).

3.2. Serotyping of L. Monocytogenes Isolates

The serovar distribution among *L. monocytogenes* isolated from poultry marketed in Iran is reported in Table 2. Three different serovars were detected: 1/2a, 1/2b and 4b. Serovar 1/2a (61%) was the predominant among *L. monocytogenes* isolates followed by serovars 1/2b and 4b (19% each).

3.3. Antimicrobial Susceptibility Test

The antimicrobial susceptibility of *L. monocytogenes* isolated from poultry marketed in Iran is showed in Table 3. Over the 21 *L. monocytogenes* strains isolated from poultry, 7 isolates (33%) were susceptible to all tested antimicrobials.

The remaining 14 isolates (77%) were found to have resistance towards one or more antimicrobials: 5 strains (24%) were resistant to 1 antimicrobial, 6 strains (29%) were resistant to 2 antimicrobials and 3 strains (14%) were resistant to ≥ 3 antimicrobials. The results of the present survey indicate a high resistance of *L. monocytogenes* to *Tetracycline* and *Pen G*, some of the antimicrobials of choice for treatment of Listeriosis. Previous studies found that *L. monocytogenes* strains isolated from poultry were resistant to *Ampicillin* and *Penicillin* (Ayaz and Erol, 2010; Fallah et al. 2012).

According to Fallah et al. (2012) the high antimicrobial resistance rate to *Tetracycline* could be explained by the frequent use of these antimicrobials to treat infection in Iranian poultry farms.

Table 2. Distribution of serovars among *L. monocytogenes* isolates

	L. monocytogenes serovars		
	1/2a	1/2b	4b
Chicken (13)	8 (61.5%)	2 (15.4%)	3 (23.1%)
Turkey (8)	5 (62.5%)	2 (25%)	1 (12.5%)
Total (21)	13 (61.9%)	4 (19%)	4 (19%)

Table 3. Antimicrobial susceptibility among *L. monocytogenes* isolates

	L. monocytogenes isolates		
	R	I	S
Tetracycline (30µg)	7	1	5
Gentamicin (10µg)	1	-	12
Chloramphenicol (30µg)	-	-	13
Vancomycin (30µg)	2	-	11
Rifampicin (5µg)	-	-	13
Pen G (10U)	6	-	7
Kanamycin (30µg)	-	-	13
Erythromycin (15µg)	3	-	10
Clindamycin (2µg)	-	-	13
Amoxicillin-clavulanic acid (20/10µg)	2	-	11

The results of this survey highlighted the high susceptibility of *L. monocytogenes* to some antimicrobials commonly used to treat Listeriosis (*Rifampin*, *Gentamycin*, *Erythromycin* and *Vancomycin*) and are in agreement with Pesavento et al. (2010), Alonso-Hernando et al. (2012), Altuntas et al. (2012) and Fallah et al. (2012).

3.4. Presence of Antimicrobial Resistance Genes by PCR Analysis

The presence of antimicrobial resistance genes was not completely correlated with antimicrobial resistance. Out of 11 strains showing resistance towards *Tetracycline*, the presence of *tet*A and *tet*B genes was found in 9 and 7 strains respectively (Figure 2A). The presence of *pen*A gene was highlighted in 87.5% of strains showing resistance towards *Pen G* (Figure 2A) All the strains showing resistance towards *Vancomycin* and *Erythromycin* presented the resistance genes *van*A/*van*B (Figure 2B and *erm*B respectively (Figure 2C).

3.5. Presence of Virulence Genes by PCR Analysis

The PCR assay highlighted that not all *L. monocytogenes* isolates carry the full complement of virulence genes (Figure 3 and Figure 4). PCR products of the nine virulence-associated genes (*inl*A, *inl*B, *inl*C, *inl*J, *act*A, *hly*A, *iap*, *plc*A, *prf*A) were obtained from 85% of the *L. monocytogenes* isolates. Lacking of *prf*A gene was highlighted in 3 strains and lacking of *inl*J and *plc*A genes was showed in 2 strains. One strain showed lacking of *iap* gene. The absence of one or more PCR products is probably due to the fact that the primers may not hybridize sufficiently to one or more allelic variants of a gene.

CONCLUSION

The main results of the present study confirmed the high prevalence of *Listeria* spp in Iranian poultry products and the high rate of resistance of *L. monocytogenes* to antimicrobials commonly used to treat Listeriosis.

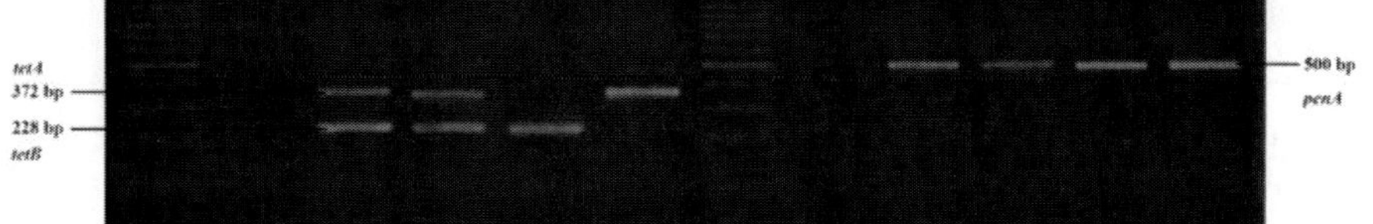

Lanes 1 and 7: 100 bp molecular size marker; lanes 3 to 6, and 9 to 12: *L. monocytogenes* isolates; lanes 2 and 8: negative control.

Figure 2A. A reprehensive gel of PCR-amplified antimicrobial resistance genes *tet*A, *tet*B, *Pen*A products: *tet*A, *tet*B, and *pen*A genes are indicated by the amplicons of 372 bp, 328 bp, and 500 bp, respectively.

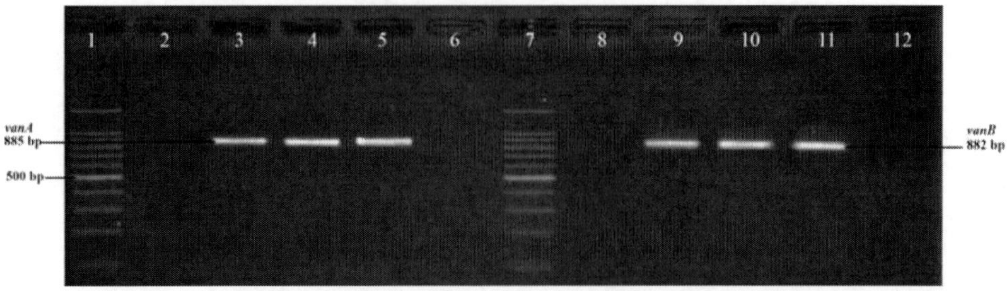

Lanes 1 and 7: 100 bp molecular size marker; lanes 3 to 6, and 9 to 12: *L. monocytogenes* isolates; lanes 2 and 8: negative control.

Figure 2B. A reprehensive gel of PCR-amplified antimicrobial resistance genes *van*A, and *van*B products: *van*A, and *van*B genes are indicated by the amplicons of 885 bp, and 882 bp, respectively.

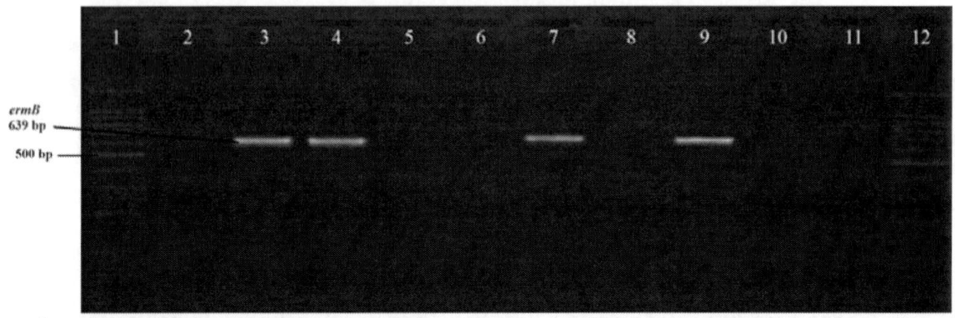

Lanes 1 and 12: 100 bp molecular size marker; lanes 2 to 10: *L. monocytogenes* isolates; lane 11: negative control.

Figure 2C. A reprehensive gel of PCR-amplified antimicrobial resistance genes *erm*B products: *erm*B gene is indicated by the amplicons of 639 bp.

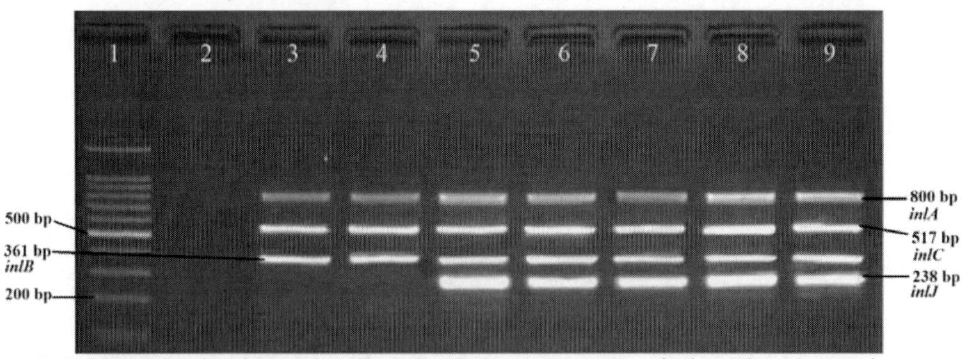

Lane 1: 100 bp molecular size marker; lanes 3 to 9: *L. monocytogenes* isolates; lane 2: negative control.

Figure 3. A reprehensive gel of PCR-amplified products of four *L. monocytogenes* virulence genes: *inl*A, *inl*B, *inl*C, and *inl*J genes are indicated by the amplicons of 800 bp, 361 bp, 517 bp, and 238 bp, respectively.

AMR is confirmed as a serious risk to public health. Although the use of antimicrobials in Iranian poultry production has reduced the risk of infectious diseases, the proper use of antimicrobials for therapeutic purposes avoiding overuse or misuse is fundamental to limit the emergence and spread of resistant *L. monocytogenes* strains in Iran.

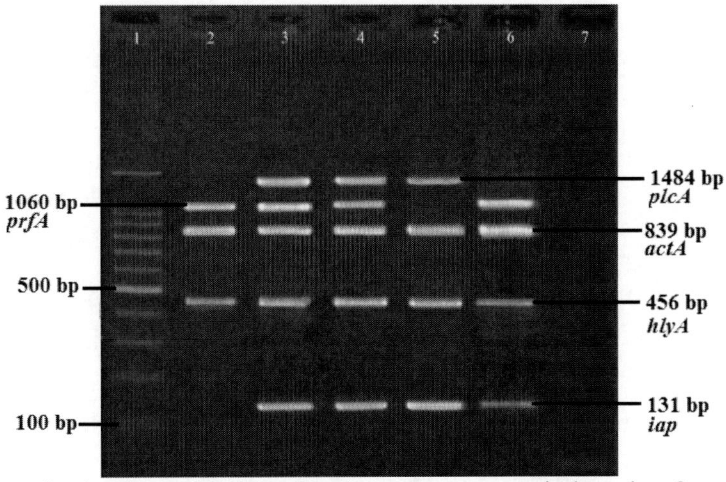

Lane 1: 100 bp molecular size marker; lanes 3 to 6: *L. monocytogenes* isolates; lane 2: negative control.

Figure 4. A reprehensive gel of PCR-amplified products of five *L. monocytogenes* virulence genes: *plc*A, *prf*A, *hly*A, iap, and *act*A genes are indicated by the amplicons of 1484 bp, 1060 bp, 456 bp, and 131 bp, and 839 bp, respectively.

REFERENCES

Alonso-Hernando, A., Prieto, M., García-Fernández, C., Alonso-Calleja, C. and Capita, R. (2012). Increase over time in the prevalence of multiple antibiotic resistance among isolates of *Listeria monocytogenes* from poultry in Spain. *Food Control*, 23, 37-41.

Altuntas, E. G., Kocan, D., Cosansu, S., Ayhan, K., Juneja, V. K. and Materon, L. (2012). Antibiotic and bacteriocin sensitivity of *Listeria monocytogenes* strains isolated from different foods. *Food and Nutrition Sciences*, 3, 363-368.

Antignac, A., Kriz, P., Tzanakaki, G., Alonso, J. M. and Taha, M. K. (2001). Polymorphism of *Neisseria meningitidis penA* gene associated with reduced susceptibility to penicillin. *J. Antimicrob. Chemother.* 47, 285-296.

Aureli, P., Fiorucci, G. C., Caroli, D., Marchiaro, G., Novara, O., Leone, L., Salmaso, S. (2000). An outbreak of febrile gastroenteritis associated with corn contaminated with Listeria monocytogenes. *New Engl. J. Med.* 2000, 342, 1236-1241.

Ayaz, N. D. and Erol, I. (2010). Relation between serotype distribution and antibiotic resistance profiles of *Listeria monocytogenes* isolated from ground Turkey. *Journal of Food Protection*, 73, 967-972.

Bailey, J. S., Fletcher, D. L. and Cox, N. A. (1989). Recovery and serotype distribution of *Listeria monocytogenes* from broiler chickens in the southeastern United States. *Journal of Food Protection*, 52, 148-150.

Clinical and Laboratory Standards Institute (CLSI). (2006). *Performance standards for antimicrobial disk susceptibility tests* (9th ed.). Wayne, PA: CLSI. Approved standard. Document M2-A9.

Clinical and Laboratory Standards Institute (CLSI). (2007). *Disk diffusion Supplemental Tables*. Wayne, PA: CLSI. Document M100-S17.

Denny, J., McLauchlin, J. (2008). Human *Listeria monocytogenes* infections in Europe-an opportunity for improved European surveillance. *Euro Surveill.* 2008, 13, 8082.

Fallah A. A., Siavash Saei-Dehkordi S., Rahnama M., Tahmasby H., Mahzounieh (2012). Prevalence and antimicrobial resistance patterns of Listeria species isolated from poultry products marketed in Iran. *Food Control* 28 (2012) 327-332.

Filiousis, G., Johansson, A., Frey, J. and Perreten, V. (2009). Prevalence, genetic diversity and antimicrobial susceptibility of *Listeria monocytogenes* isolated from open-air food markets in Greece. *Food Control*, 20, 314-317.

Gebreyes, W. A. and Altier, C. (2002). Molecular characterization of multidrug-resistant *Salmonella enterica* subsp. *enterica* serovar Typhimurium isolates from swine. *J. Clin. Microbiol.* 40, 2813-2822.

Gillespie, I. A., McLauchlin, J., Grant, K., Little, C. L., Mithani, V., Penman, C., Lane, X. C., Regan, M. (2006). Changing pattern of human listerios, England and Wales, 2001-2004. *Emerg. Infect. Dis.* 2006, 12, 1361-1366.

Gudbjörnsdóttir, B., Suihko, M. L., Gustavsson, P., Thorkelsson, G., Salo, S., Sjöberg, A. M., (2004). The incidence of *Listeria monocytogenes* in meat, poultry and seafood plants in the Nordic countries. *Food Microbiology*, 21, 217-225.

Guillaume, G., Verbrugge, D., Chasseur-Libotte, M. L., William, M. and Collard, J. M. (2000). PCR typing of tetracycline resistance determinants (Tet A-E) in *Salmonella enterica* serotype Hadar and in the microbial community of activated sludges from hospital and urban wastewater treatment facilities in Belgium. *FEMS Microbiol. Ecol.* 32: 77-85.

Ganasena, D. K., Kodikara, C. P., Ganepola, K. and Widanapathirana, S. (1995). Occurrence of *Listeria monocytogenes* in food in Sri Lanka. *Journal of the National Science Council of Sri Lanka*, 23, 107-114.

ICMSF (International Commission on Microbiological Specifications for Foods) (1998). Microorganisms in foods. In: *Microbial specifications of food commodities*, Vol 6. London: Blackie Academic and Professional.

Jamali, H. and Thong, K. L. (2014). Genotypic characterization and antimicrobial resistance of *Listeria monocytogenes* from ready-to-eat foods. *Food Control*, 44, 1-6.

Krueger, A. L., Greene, S. A., Barzilay, E. J., (2014). Clinical outcomes of nalidixic acid, ceftriaxone, and multidrug resistant nontyphoidal Salmonella infections compared with pansusceptible infections in FoodNet sites, 2006-2008 External Web Site Icon. *Foodborne Pathog. Dis.* 2014; 11(5): 335-41.

Lanz, R., Kuhnert, P. and Boerlin, P. (2003). Antimicrobial resistance and resistance gene determinants in clinical *Escherichia coli* from different animal species in Switzerland. *Vet. Microbiol.* 91, 73-84.

Loura, C. A. C., Almeida, R. C. C. and Almeida, P. F. (2005). The incidence and level of Listeria spp. and *Listeria monocytogenes* contamination in processed poultry at a poultry processing plant. *Journal of Food Safety*, 25, 19-29.

McClain, D. and Lee, W. H. (1988). Development of USDA-FSIS method for isolation of *Listeria monocytogenes* from raw meat and poultry. *Journal of the Association of Official Analytical Chemists*, 71, 660-664.

Meloni, D. (2014). Focusing on the main morphological and physiological characteristics of the food-borne pathogen Listeria monocytogenes. *J. Vet. Sci. Res.* 2014, 1, 1-2.

Osaili, T. M., Alaboudi, A. R. and Nesiar, E. A. (2011). Prevalence of Listeria spp. and antibiotic susceptibility of *Listeria monocytogenes* isolated from raw chicken and ready-to-eat chicken products in Jordan. *Food Control*, 22, 586-590.

Pesavento, G., Ducci, B., Nieri, D., Comodo, N. and Lo Nostro, A. (2010). Prevalence and antibiotic susceptibility of Listeria spp. isolated from raw meat and retail foods. *Food Control*, 21, 708-713.

Public Health Agency Canada (2008). *Lessons Learned: Public Health Agency of Canada's Recall Response to the 2008 Listeriosis Outbreak.* Available online: http://www.phac-aspc.gc.ca/fs-sa/listeria/2008-lessons-lecons-eng. php (accessed on 4 February 2015).

Rawool, D. B., Malik, S. V. S., Shakuntala, I., Sahare, A. M. and Barbuddhe, S. B. (2007). Detection of multiple virulence-associated genes in *Listeria monocytogenes* isolated from bovine mastitis cases. *International journal of food microbiology*, 113, 201-207.

Romanova, N., Favrin, S., Griffiths, M. W. (2002). Sensitivity of *Listeria monocytogenes* to sanitizers used in the meat processing industry. *Appl. Environ. Microbiol.* 2002, 68, 6405-6409.

Rossmanith, P., Krassnig, M., Wagner, M. and Hein, I. (2006). Detection of *Listeria monocytogenes* in food using a combined enrichment/real-time PCR method targeting the *prfA* gene. *Research in Microbiology*, 157, 763-771.

Sakaridis, I., Soultos, N., Iossifidou, E., Papa, A., Ambrosiadis, I. and Koidis, P. (2011). Prevalence and antimicrobial resistance of *Listeria monocytogenes* isolated in chicken slaughterhouses in northern Greece. *Journal of Food Protection*, 74, 1017-1021.

Shen, Y., Liu, Y., Zhang, Y., Cripe, J., Conway, W., Meng, J., Hall, G., Bhagwat, A. A. (2006). Isolation and characterization of *Listeria monocytogenes* isolates from ready-to-eat foods in Florida. *Appl. Environ. Microbiol.* 2006, 72, 5073-5076.

Skovgaard, N. and Morgen, C. A. (1988). Detection of Listeria spp. in faeces from animals, in feeds, and in raw foods of animal origin. *International Journal of Food Microbiology*, 6, 229-242.

Todd, E. C. D., Notermans, S. Surveillance of listeriosis and its causative pathogen, Listeria monocytogenes. *Food Control* 2011, 22, 1484-1490.

Ueda, F., Sugamata, M., Aota, M., Mochizuki, M., Yamada, F. and Hondo, R. (2002). Swift and definite serotyping for isolated *Listeria monocytogenes* strains. *Microbiologica* 25, 165-171.

Uyttendale, M. R., Neyts, K. D., Lips, R. M. and Devebere, J. M. (1997). Incidence of *Listeria monocytogenes* in poultry products obtained from Belgian and French abbatoirs. *Food Microbiology*, 14, 339-345.

US Department of Agriculture's Food Safety and Inspection Service (USDA/FSIS) (2010). Interpretive Summary: Quantitative Assessment of Relative Risk to Public Health from Foodborne *Listeria monocytogenes* among Selected Categories of Ready-to-Eat Foods. Available online: http://www.fda.gov/downloads/food/foodscienceresearch/ucm197329. pdf (accessed on 4 February 2015).

Van Coillie, E., Werbrouck, H., Heyndrickx, M., Herman, L., Rjipens, N. (2004). Prevalence and typing of *Listeria monocytogenes* in Ready-to Eat food products on the Belgian market. *J. Food Prot.* 2004, 67, 2480-2487.

Yücel, N., Çitak, S. and Önder, M. (2005). Prevalence and antibiotic resistance of Listeria species in meat products in Ankara, Turkey. *Food Microbiology*, 22, 241e245.

In: Listeria monocytogenes
Editor: Tiffany Vicario

ISBN: 978-1-63483-765-1

Chapter 6

ADHESION AND BIOFILM FORMATION OF *LISTERIA MONOCYTOGENES* ON FOOD CONTACT SURFACES: GENERAL ASPECTS AND CONTROL MEASURES

Letícia Sopeña Casarin[1,2,*] and Eduardo Cesar Tondo[1]
[1]Instituto de Ciência e Tecnologia de Alimentos, Universidade Federal do Rio Grande do Sul – ICTA/UFRGS, Brazil
[2]Departamento de Nutrição, Universidade Federal de Ciências da Saúde de Porto Alegre UFCSPA, Brazil

ABSTRACT

Currently *Listeria monocytogenes* is one of the most important food pathogens worldwide. This saprophytic microorganism is responsible for a highly fatal foodborne disease in humans and can be found inside food industries, mainly because of poor hygiene practices, unhygienic design of equipment and cross-contamination. Several studies have shown that *L. monocytogenes* is capable to attach and form biofilms on the surfaces of many materials, as metal, glass, polyethylene, and rubber, being able to persist for long periods in food plants. It has been shown that *L. monocytogenes* strains that persist in food processing environments form thicker biofilms than planktonic strains, indicating that the biofilms are an important strategy for the survival of these bacteria. For example, bacterial cells within biofilms are much more resistant to antimicrobials, detergents, and sanitizing agents than their planktonic counterparts. However, the biofilm formation and its preventive measures depends on several aspects, such as the lineage of *L. monocytogenes*, its origin of isolation, nutrient levels present in media, environmental temperature, and hydrophobicity of surfaces and cells. The purpose of this review is to study the general aspects and how to control the *L. monocytogenes* biofilms on food contact surfaces.

1. General Aspects of Listeria and Listeriosis

Listeria is a Gram-positive, facultative anaerobic, highly mobile, rod-type, and non-spore forming bacterium (Setiani et al. 2015). The genus *Listeria* consists of fifteen closely related species: *L. innocua, L. ivanovii, L. marthii, L. monocytogenes, L. murrayi* (also called *L. grayi), L. rocourtiae, L. seeligeri, L. welshimeri, L. fleiscmannii, L. weihenstephanensis, L. floridensis, L. aquatic, L. cornellensis, L. riparia, and L. grandensis* (Weller et al. 2015; Den Bakker et al. 2014; Lang Halter, Neuhaus and Scherer, 2013; Leclercq et al. 2010). In addition, some novel species have been proposed (Weller et al. 2015). According to the literature, only *L. monocytogenes* has been commonly associated with human listeriosis, because listeriosis caused by *L. ivanovii, L. welshimeri,* or *L. seeligeri* are extremely rare in humans. *L. innocua* and *L. grayi*, and the others are nonpathogenic until now (Setiani et al. 2015; Weller et al. 2015).

In 1980, *L. monocytogenes* was recognized as an emerging foodborne pathogen because was responsible for several sporadic cases and foodborne disease outbreaks. This microorganism was responsible for 30% of all deaths attributed to foodborne pathogens from 1996 to 2005 in the United States (Barton Behravesh et al. 2011), being currently one of the most important food pathogens with expressive impact on public health and economy worldwide (Ebner et al. 2015).

L. monocytogenes is the etiological agent of human and animal listeriosis, a highly fatal infection associated with the ingestion of contaminated food (Mackiw et al. 2016). During the last three decades, a large number of listeriosis outbreaks and sporadic cases of varied extent have been reported in North America and Europe, and diverse food products have been involved with the transmission of listeriosis.

Several outbreaks have been associated with the consumption of dairy products, particularly soft cheese, under-cooked meat, fish, eggs and egg products, ready-to-eat products, unwashed raw vegetables and cabbage (Alonso-Hernando et al. 2012; Bouayad and Hamdi, 2012; Cokal et al. 2012; Jakobsen et al. 2011; Rahimi, Ameri and Momtaz, 2010; Garrido, Vitas and Gracia-Jalon, 2009; Filiousis et al. 2009; Harakeh et al. 2009; Arslan and €Ozdemir, 2008; de Santis et al. 2007; Gibbons et al. 2006; Aureli et al. 2003).

In 2013, European countries reported 1763 confirmed human cases of listeriosis, which indicated a 7.3% increase compared with 2012. The European Union notification rate was 0.44 cases per 100,000, and the highest specific notification rate was observed in Finland (1.12), while the lowest in Bulgaria (0.04) (EFSA, 2015).

Although listeriosis has a low incidence compared to other foodborne diseases such as salmonellosis or campylobacteriosis, this disease is recognized as a major issue to public health authorities due to its high hospitalization rate (94%) and high case-fatality rate (12.8 to 17% of cases) (ECDC, 2013; CDC, 2011). Fatality rates of listeriosis exceed those of even *Salmonella* spp. and *Clostridium botulinum* (Lee et al. 2015).

Clinical manifestations of invasive listeriosis are usually severe, including abortion, sepsis, and meningoencephalitis. Listeriosis can also manifest as a febrile gastroenteritis syndrome (Vazquez-Boland et al. 2001). *L. monocytogenes* has been differentiated into 13 serotypes 1/2a, 1/2b, 1/2c, 3a, 3b, 3c, 4a, 4ab, 4b, 4c, 4d, 4e, and 7 (; Jamali, Radmehr and Thong, 2013; Liu, 2006). Molecular serogroup-related PCR typing of *L. monocytogenes* is based on the detection of serogroupspecific regions, leading to five genosero groups

(Lineage): IIa (serotype 1/2a-3a), IIb (serotypes 1/2b-3b-7), IIc (serotypes 1/2c-3c), IVa (serotype 4a-4c), IVb (serotypes 4ab-4b, 4d-4e) (Doumith et al. 2004; Borucki and Call, 2003).

Of the serotypes of *L. monocytogenes* which can cause illness, only four of them account for 95% of human isolates: 1/2a, 1/2b, 1/2c and 4b (Huang et al. 2011; Martins and Leal Germano, 2011; Uhitil et al. 2004; Kathariou, 2002). In the US, in 2012, serotype 4b was the most commonly identified serotype (54%) and serotype 1/2a was the second (28%) (CDC, 2014). The most frequently isolated serotype from clinical samples in Brazil was 4b (60.3%), followed by 1/2a (29.0%) (Hofer, Reis and Hofer, 2006). Serotype 1/2a has frequently been detected in different food matrices (Jamali, Radmehr and Thong, 2013; Martins and Leal Germano, 2011). In Europe during 2013, the most common serotypes isolated from foodborne outbreaks samples were 1/2a (57.5%) and 4b (34.3%), followed by 1/2b (6.4%), 1/2c (1.4%), 3a and 3b (both 0.2%). The most common Lineage identified was IIa (corresponding to serotypes 1/2a and 3a), followed by IVb (serotypes 4b, 4d, and 4e), IIb (serotypes 1/2b, 3b), and IIc (serotypes 1/2c and 3c) (EFSA and ECDC, 2014). A possible correlation between serotype or lineage and biofilm forming capacity is still unclear.

L. monocytogenes is of prime interest due to its ability to survive and grow in a wide range of environmental conditions, including refrigeration temperatures (Lee et al. 2015). It can grow under temperatures between -1.5°C to 45°C, and at pH range between 4.4 and 9.4, with an optimum pH 7 (Setiani et al. 2015). The robustness and ubiquity of *L. monocytogenes* in the environment make this microorganism a post processing contamination hazard in the food industry (Tompkin, 2002).

Listeria spp., including *L. monocytogenes,* have been isolated from the environment of various types of food processing facilities, e.g., meat, poultry, dairy, and seafood processing facilities (Williams et al. 2011; Eifert et al. 2005; Lappi et al. 2004). Control of *L. monocytogenes* is a continuous challenge for the food industry (Ferreira et al. 2011), mainly because it is easily spread by water, air flow, utensils, humans, among others.

In response to adverse conditions, *L. monocytogenes* can develop biofilms either on abiotic or biotic surfaces, resulting in persistence on food-processing surfaces and subsequent cross-contamination or re-contamination of food products (Renier, Hébraud and Desvaux, 2011; Tresse et al. 2007; Beresford, Andrew and Shama, 2001).

2. Adhesion and Biofilm Formation of *L. monocytogenes* on Food Contact Surfaces

Biofilms are described as surface adhering bacterial cells organized in clusters and enclosed in self-produced extracellular polymeric substances (EPS), mainly composed of polysaccharides and proteins (Djordjevic, Wiedmann and McLandsborough, 2002; Watnick and Kolter, 2000; Costerton et al. 1995; Costerton et al. 1994). The biofilm formation results from a continuous process of adhesion to a surface, development of a bacterial community structure and detachment off cells. The initial step for biofilm formation is the bacterial adhesion to the surface (Takhistov and George, 2004). It starts with cell surface contact that triggers a reversible adhesion of the two corpuses through van der Waals, electrostatic and Lewis acid base interactions (Bellon-Fontaine et al. 1990).

In the case of *L. monocytogenes*, the electrostatic charge of bacterial cell walls (conferred by peptidoglycan anionic teichoic acids) and cell surface hydrophobicity (enhanced by the presence of lactic acid) have been shown to govern its attachment to stainless steel (Briandet et al. 1999a; 1999b). This reversible adhesion becomes irreversible through stronger bonds, leading to the initiation of biofilm formation (Marshall, 1992; van Loosdrecht et al. 1989). The presence of proteins anchored to the cell wall may drastically affect microbial behavior at the interfaces (Bridier et al. 2015). In *L. monocytogenes*, the inactivation of the SecA2 protein export pathway has shown to impact biofilm formation and architecture (Renier et al. 2013). Curiously, an ubiquitous protein cell wall anchoring system mediated by transpeptidases (denominated sortases) was characterized in Gram-positive bacteria (Mazmanian et al. 1999). This system allows the covalent linkage of cell surface proteins harboring an LPXTG amino-acid signature to the cell wall, some of them being involved in both adhesion and biofilm development (Guiton et al. 2009).

The cell attachment to the surface is influenced by physico-chemical properties of the materials including charge, liquid velocity (static/dynamic flow biofilms), roughness, hydrophobicity (nonpolar surface such as teflon and other plastics), and hydrophilicity (glass or metals) (Donlan, 2002; Kumar and Anand, 1998). The surfaces of most bacterial cells are negatively charged, and the negative charge of the cell surface is adverse to bacterial adhesion, due to electrostatic repulsive force. However, the bacterial cell-surface is hydrophobic due to fimbriae, flagella and lipopolysaccharide (LPS) (Takahashi et al. 2010; Di Bonaventura et al. 2008). Hydrophobic interactions between the cell surface and the substratum may enable the cell to overcome repulsive forces and attach (Donlan, 2002). Even though diverse studies have been published about bacterial attachment (Bonsaglia et al. 2014; Casarin et al. 2014; Szlavik et al. 2012; Silva et al. 2008), the role of hydrophobicity of the cell and surfaces, and the role of surface roughness on bacterial adhesion is still controversial. Our studies demonstrated that the number of attached *L. monocytogenes* on stainless steel and MIG and TIG welds was approximately 3.0 $\log_{10}$ CFU/cm^2, after 8h of contact time, with an initial inoculum of 5.0 $\log_{10}$ (Casarin et al. 2014).

Furthermore, there was no difference between the adhesions of this microorganism in different welds and the welded and not welded 304 and 316 stainless steel surfaces. Bacterial attachment was not influenced by roughness of MIG and TIG welding (Casarin et al. 2014). These findings demonstrated that, contrary to our expectations, the surface roughness did not influence in the *L. monocytogenes* adhesion (Casarin et al. 2014). Similar results were found by Tide et al. (1999) and Flint (1996).

The biofilm growth is mainly characterized by aggregate formation and EPS secretion by cells, while the biofilm maturation is reached when these 3-D structures are crossed by water channels or pores which ensure nutrients exchange and removal of bacterial waste products (Sauer, Rickard and Davies, 2007; van Loosdrecht et al. 1989). The bacteria involved in the biofilm formation undergo transitions from planktonic and sessile forms and the net accumulation of biomass in a mature biofilm depends on cell growth and cell detachment (Pilchová et al. 2014). Mushroom-shaped, honeycomb-like and ball-shaped structures are the main architectures described for listerial biofilms (Rieu et al. 2008a; Hall-Stoodley and Stoodley, 2002). A broad diversity of extracellular appendages such as flagella, pili may be involved in the initial stages of biofilm formation (Van Houdt and Michiels, 2010). Flagella are not only the principal components for bacterial motility, but also play an important role in cell-surface and cell-cell interactions.

Flagella have been shown to be critical to the initial contact and structuring of *E. coli* and *Yersinia enterocolitica* biofilms (Kim et al. 2008; Van Houdt and Michiels, 2005; Pratt and Kolter, 1998). A flagella-deficient mutant of *L. monocytogenes* displayed reduced initial adhesion but hyper-biofilm development under flow conditions (Todhanakasem and Young, 2008). In Gram-positive bacteria pili have been definitely assigned as important players for the initial adhesion of bacteria to surfaces, in bacterial aggregation, and in biofilm formation (Mandlik et al. 2008). This was demonstrated for *Enterococcus* spp. (Garsin and Willems, 2010; Sillanpaa et al. 2010).

The colonization of surfaces by *L. monocytogenes* depends on environmental conditions, as well, including the type of medium, temperature, pH, hydrodynamic conditions, and the presence of other bacteria able to biofilm formation (Di Bonaventura et al. 2008; Tresse et al. 2007; Djordjevic et al. 2002). For instance, the presence or the motility of flagella plays a predominant role, at least in the early stages of biofilm formation. The temperature regulates the formation of flagella on *L. monocytogenes* cells (van Houdt and Michiels, 2010; Vatanyoopaisarn et al. 2000; Todhanakasem and Young, 2008; Lemon, Higgins and Kolter, 2007), and, consequently, temperature effect has been widely investigated in biofilm formation of *L. monocytogenes* (Vatanyoopaisarn et al. 2000; Peel, Donachie and Shaw, 1988; Moens and Vanderleyden, 1996). Kadam et al. (2013) studied the biofilm formation at 12°C, 20°C, 30°C, and 37°C and demonstrated that the biofilm formation was reduced when temperatures were decreased. Tresse et al. (2007) have reported pH dependence for *L. monocytogenes* flagellation and consequences on cell attachment.

The complex matrix of mature biofilms acts as a barrier against environmental stresses and antimicrobial agents, which can be considered adequate strategy for bacterial survival. The principal components of this gelatinous material are water and biopolymers (Flemming, 2011; Flemming and Wingender, 2010). Extracellular DNA has been more recently identified as a contributor to the matrix and has been detected in large quantities in both Gram-negative and Gram-positive biofilms (Izano et al. 2008). Studies have shown that foodborne pathogens such as *L. monocytogenes* and *B. cereus* release DNA as an adhesion mediator or a major component in their biofilm matrix (Harmsen et al. 2010; Vilain et al. 2009). Some reports have described proteinaceous amyloid fibers as a common structural component in biofilm matrices (Dueholm et al. 2010; Romero et al. 2010).

The different stages of surface contamination are driven by quorum-sensing (QS) and other communication strategies using genetic regulators (Bridier et al. 2015). The accessory gene regulator (agr) system of *L. monocytogenes* plays a central role in QS signaling, because it is essential to the complex 3D architectures and mature biofilm phenotypes (Rieu et al. 2007; Autret et al. 2003). Rieu et al. (2008b) demonstrated that the in-frame deletion of *agrA*, which encodes a transcriptional regulator, resulted in a decrease in the initial adherence of *L. monocytogenes* on AISI 304 stainless steel surface, without affecting the subsequent biofilm development. However, its complex role in biofilm formation is probably strain-and growth condition-dependent (Yarwood et al. 2004).

A possible correlation between serotype or lineage and biofilm forming capacity is still controversial. Lineage I strains showed to produce more biofilm than strains from lineage II (Djordjevic et al. 2002), while others researches showed that 1/2a and/or 1/2c strains (lineage II) were better biofilm formers than 4b strains (lineage I) (Combrouse et al. 2013; Nilsson, Ross and Bowman, 2011; Pan, Breidt and Gorski, 2010; Harvey, Keenan and Gilmour, 2007; Borucki et al. 2003).

Kadam et al. (2013) demonstrated that the type of serotype had a significant effect on biofilm formation and was influenced by medium tested, because strains from both serotypes 1/2b and 1/2a formed more biofilm than strains of serotype 4b in nutrient-rich medium at 20°C, 30°C and 37°C. In nutrient-poor medium, the biofilm production levels demonstrated by strains of serotypes 1/2a and 4b were rather similar and lower than strains of serotype 1/2b (Kadam et al. 2013).

The origin of isolation of strains might affect biofilm formation (Nilsson, Ross and Bowman, 2011). However, Kadam et al. (2013) observed that the level of biofilm formation was not significantly affected by the origin of isolation (including dairy, meat, industrial environment, human and animal), irrespective of medium used and temperature tested; so the influence of origin of isolation in the biofilm formation remains unclear.

3. Advantages of Biofilm Formation As Survival Strategies of *L. monocytogenes*

Biofilm formation provides different advantages to bacterial cells, which can be considered excellent survival strategies. Biofilm formation represents at least four interconnected advantages for bacterial cells: (i) protection from stressful environmental conditions, by providing a certain degree of shelter and homeostasis, (ii) benefits of metabolic interactions among microbial species due to commensalism, cooperation to mutualism, (iii) competition for and appropriation of available nutrients, and (iv) gene transfer enabling acquisition of new adaptative phenotypic traits (Giaours et al. 2014; Molin and Tolker-Nielsen 2003; Davey and O' Toole 2000).

Cells in biofilms present increased resistance to disinfection procedures, compared to planktonic cells, resulting in increased survival. There are at least four mechanisms for explaining the increased resistance to sanitizers of cells in biofilms (Donlan and Costerton, 2002; Gilbert, Allison and McBain, 2002): (i) the exopolymeric matrix forms a physical barrier that limits the diffusion of sanitizers within the biofilm, (ii) resistance mechanisms encoded by specific genes or clusters of genes can be shared among cells (e.g., detoxifying membrane transporters). These genes can be present in plasmids and can be horizontally transferred among biofilm cells, (iii) the differentiation of bacterial cells into different physiological states (e.g., dormant cells) less susceptible/receptive to treatments, and (iv) the modification of the microenvironment (e.g., local acidic pH), rending a particular sanitizer less efficient (Giaours et al. 2014). *Listeria* can either exist in monoculture biofilms or be a part of mixed culture biofilms, with bacteria such as *Flavobacterium* (Bremer, Monk and Osborne, 2001). Bremer, Monk and Osborne (2001) investigated the attachment levels of *L. monocytogenes* to stainless steel surfaces, when the microorganism was cultivated in a pure culture and also with *Flavobacterium*. Researchers found that significantly higher numbers of *L. monocytogenes* cells attached to stainless steel surfaces when *Flavobacterium* was present.

Further, *L. monocytogenes* cells survived longer periods of time when was in this mixed culture (Bremer, Monk and Osborne, 2001). An increasing number of studies have reported that multispecies biofilms appear to be more resistant to antimicrobial activity than their mono-species counterparts (Giaouris et al. 2013; Schwering et al. 2013; van der Veen and Abee, 2011; Luppens et al. 2008; Burmolle et al. 2006).

For instance, it was shown that under most conditions, *L. monocytogenes* in a mixed biofilm with *Lactobacillus plantarum* exhibited higher resistance to benzalkonium chloride and peracetic acid than single species biofilms (van der Veen and Abee, 2011). This enhanced resistance was linked to the protection offered by the resistant species to the whole community, rather than selection for the resistant species (Bridier et al. 2015). It is important to emphasize that the resistance to sanitizers may cause the persistence of *L. monocytogenes* in food processing environment for years (Carpentier and Cerf, 2011).

Some observations indicate that the spatial organization of species, and the sharing of matrix components, may therefore determine the degree of persistence of pathogens in mixed communities as well as their resistance to cleaning and disinfection treatments (Bridier et al. 2015). In addition to such unspecific processes, some findings have suggested that microbial species may actively cooperate. Xavier and Foster (2007) showed that the presence of multiple strains in a biofilm can promote the production of extra-polymeric substances, which are key components for the persistence of biofilms on a surface, and its resistance. Throughout this communication and cooperation among different cells and species, a biofilm thus offers an environment that can promote the emergence and dissemination of specific functional properties that will safeguard the community in the face of adverse conditions (Sachs and Hollowell, 2012).

An important finding is that persistent strains showed higher tolerance to acid, and lower to heat, when compared to non-persistent strains (Lundén et al. 2003). The persistence of *L. monocytogenes* strains in food processing has also been associated with its ability to tolerate desiccation when attached to stainless steel (Vogel et al. 2010). Strains of *L. monocytogenes* were able to survive for extended periods of time on dry stainless steel surfaces and some physical and chemical parameters in this environment seem to affect this ability. The presence of organic material and particularly sodium chloride enhanced significantly the survival of strains, which remained viable after extended periods of dry conditions (Vogel et al. 2010). It has been shown that sodium chloride can cause aggregation and adherence of *L. monocytogenes* to plastic surfaces (Jensen et al. 2007) and this process may have a protective effect on the cells, suggesting that stress responses to salt and desiccation may overlap (Melo, Andrewb and Faleiro, 2015). All these characteristics may have an effect on the initiation of a persistence process in a food processing plant. Indeed, persistent contamination is not due to only one single factor but rather the sum of several interacting factors (Melo, Andrewb and Faleiro, 2015). Based on these observations, it would seem that targeting the mechanisms by which species interact in mixed communities might be a promising solution to control biofilms (Boyle et al. 2013), but it requires a clearer understanding of the nature of these interactions (Bridier et al. 2015). Modeling approaches focused on species interactions within biofilms have emerged, and should contribute to developing a comprehensive view of these complex microbial ecosystems (Xavier et al. 2005).

4. Control of Adhesion and Biofilm Formation

Since biofilms are a great concern in the food industries, many studies have been done in order to gain a better understanding of their development, spread, and also to come up with countermeasures (Srey, Jahid and Ha 2013).

The first and most important thing to do inside food industries is to prevent biofilm formation by regularly cleaning and sanitizing food contact surfaces. These procedures will prevent cell to firmly attach (reversible attachment) (Simões et al. 2006; Midelet and Carpentier 2004). Meyer (2003) suggested three different strategies: (i) disinfection "in time," before biofilm develops, (ii) disinfection of biofilms using harsh disinfectants, and (iii) inhibition of attachment of microbes by selecting surface materials that do not promote attachment or by supplementing with nutrients.

Even though, the regular cleaning and disinfection procedures remain the most common strategy to control pathogens attachment on industrial equipment (Jahid and Ha, 2012), they may be not fully effective to remove mature biofilms. The use of sanitizers to remove biofilms can induce the selection of resistant phenotypes, as well (Simoes, Simoes and Vieira, 2010). Moreover, the increasing concern over environmental impacts has led to the exploration of sanitizers that do not form hazardous by-products. Compounds as ozone or acidic electrolysed water are generally considered as green (or eco-friendly) biocides and they are chemical residue-free (Ashraf et al. 2014). Natural compounds extracted from bacterial cultures or aromatic plants, and "generally recognized as safe," are now being evaluated for their potential to eradicate biofilms. They may exert a high level of lethal activity against pathogens, be efficient in penetrating the structure of a biofilm, and be easily degraded in the environment (Bridier et al. 2015; Neyret et al. 2014).

The use of enzyme-based detergents are also being used once may offer additional tools to improve disinfection processes (Bridier et al. 2015). Enzymes can target and loosen the biofilm matrix, as well as being able to trigger cell release from biofilms (Bridier et al. 2015). This allows conventional disinfection agents to reach their bacterial targets embedded in a biofilm or dispersed as planktonic cells. Depending on the composition of the biofilm matrix, different enzymes can be preferred, including proteases, cellulases, DNAses, alginate lyases, polysaccharide depolymerases, or dispersin B (Jabbouri and Sadovskaya, 2010; Lequette et al. 2010; Orgaz, Neufeld and SanJose, 2007). Bacteriophages have also been appointed as potential candidates to attack pathogens in bacterial biofilms. They can diffuse through the biofilm matrix while retaining their antibacterial efficacy (Donlan, 2009; Briandet et al. 2008). Natural and engineered phages were found to be effective against food-borne pathogen biofilms (Simoes, Simoes and Vieira, 2010). Phages were found to be effective in controlling *L. monocytogenes* on stainless steel which was either clean or had been soiled with fish proteins. However, their efficacy was greater on cells dislodged from the surface than on those organized within a biofilm (Lu and Collins, 2007). Since it the eradication of the biofilm has several limitations, modifications to inert surfaces so as to reduce their interactions with microbial cells and subsequent adhesion, have been proposed (Simoes, Simoes and Vieira, 2010). In this regard, modified highly hydrophobic polystyrene demonstrated their antibacterial efficacy against different micro-organisms, including *L. monocytogenes* (Poncin-Epaillard et al. 2013). Functionalized carbohydrates acting as lectin inhibitors have been used to coat materials and prevent the production of microbial adhesins or their interactions with the surfaces (Korea, Ghigo and Beloin, 2011). The biosurfactants have also been considered as promising anti-adhesive and anti-microbial coating molecules (Meylheuc et al. 2006). The use of chemical or natural antimicrobials as alternative coating molecules has also been found to be effective in preventing both adhesion and bacterial development on food surfaces (Duan et al. 2007; Zhang et al. 2006).

Surface modifications using nano-materials, such as silver, cobalt and iron mixed oxides also have been studied (Ashraf et al. 2014). Another promising way to prevent biofilm formation is to use Quorum Sensing antagonistic molecules (Bridier et al. 2015). These molecules prevent the biofilm implementation and maturation, since interfere with bacterial communication. Compounds such as brominated furanones were shown to successfully reduce biofilm formation (Ni et al. 2009).

Finally, the last strategy, uses the concept of guided microbial ecology, protective biofilms have been proposed as barrier microflora against the potential establishment of foodborne pathogens. One example is the use of the ecology of wooden shelves used for cheese ripening (Mariani et al. 2007).

The resident microflora on these shelves enabled inhibition of *L. monocytogenes* by up to 2 log^{10} (CFU/cm^2) during 12 days of ripening, while inactivation of the autochthonous microflora enabled an increase in the *L. monocytogenes* population by 4 log (CFU/cm^2) during the same period (Mariani et al. 2007). Habimana et al. (2011) in competition studies between *Lactococcus lactis* (food grade) and *L. monocytogenes* in a dual species biofilms revealed that the shorter generation time of *L. lactis* than that of *L. monocytogenes* was the key factor involved in inhibition of the pathogen.

As for competitive exclusion, the genetic background of the protective biofilm of *Lc. lactis*, which impacts both its physiology and its physicochemical surface properties, was shown to be critical to its ability to prevent the fixation of *L. monocytogenes* (Habimana et al. 2009). Other factors as bacteriocins which are produced by such protective biofilms also display important bacterial features that may potentialize the competitive exclusion or the displacement of unwanted pathogens (Winkelströter et al. 2011; Leriche, Chassaing and Carpentier 1999).

CONCLUSION

L. monocytogenes is a saprophytic microorganism that emerged as one of the most important food pathogen around the world. Among the characteristics that allowed such importance is the ability of forming biofilms on food contact surfaces, what frequently result in food contamination and fatal human cases.

In view of the strong impact of biofilm formation on the environmental persistence and spread of *L. monocytogenes*, the knowledge about factors influencing biofilm formation is essential for its control. The present study demonstrated that food industry-persistent *L. monocytogenes* produce thicker biofilms than planktonic cells and the biofilms formed by this microorganism may be influenced by the temperature, pH and composition of media where *L. monocytogenes* is growing. The characteristics of specific strains also are important for biofilm formation, because certain pili and flagella influence positively the biofilm formation.

Even though *L. monocytogenes* biofilms represent an important risk to the public health, the most common control measure is the cleaning and sanitation of food contact surfaces. Beyond that, other strategies as the use of quorum sensing antagonistic molecules, surface modifications with nano-materials, such as silver, cobalt and iron, use of biosurfactants, bacteriophages and competitive exclusion using acid lactic bacteria has been used.

REFERENCES

Alonso-Hernando, A., Prieto, M., García-Fernandez, C., Alonso-Calleja, C., Capita, R. (2012). Increase over time in the prevalence of multiple antibiotic resistances among isolates of *Listeria monocytogenes* from poultry in Spain. *Food Control*, 23(1), 37-41.

Arslan, S. and €Ozdemir, F. (2008). Prevalence and antimicrobial resistance of *Listeria* spp. in homemade white cheese. *Food Control*, 19(4), 360-363.

Ashraf, M. A., Ullah, S., Ahmad, I., Qureshi, A. K., Balkhair, K. S., Abdur Rehman, M. (2014). Green biocides, a promising technology: current and future applications to industry and industrial processes. *Journal of the Science of Food and Agriculture*, 94, 388-403.

Aureli, P., Ferrini, A. M., Mannoni, V., Hodzic, S., Wedell-Weergaard, C., Oliva, B. (2003). Susceptibility of *Listeria monocytogenes* isolated from food in Italy to antibiotics. *International Journal of Food Microbiology*, 83(3), 325-330.

Autret, N., Raynaud, C., Dubail, I., Berche, P., Charbit, A. (2003). Identification of the agr locus of *Listeria monocytogenes*: role in bacterial virulence. *Infection and Immunity*, 71, 4463-4471.

Barton Behravesh, C., Jones, T. F., Vugia, D. J., Long, C., Marcus, R., Smith, K., Thomas, S., Zansky, S., Fullerton, K. E., Henao, O. L., Scallan, E. (2011). FoodNet Working Group Deaths associated with bacterial pathogens transmitted commonly through food: foodborne diseases active surveillance network (FoodNet), 1996-2005. *The Journal of Infectious Diseases*, 204(2), 263-267.

Bellon-Fontaine, M. N., Mozes, N., van der Mei, H. C., Sjollema, J., Cerf, O., Rouxhet, P. G., Busscher, H. J. (1990). A comparison of thermodynamic approaches to predict the adhesion of dairy microorganisms to solid substrata. *Cell Biophysics*, 17, 93-106.

Beresford, M. R., Andrew, P. W., Shama, G. (2001). *Listeria monocytogenes* adheres to many materials found in food-processing environments. *Journal of Applied Microbiology*, 90 (6), 1000-1005.

Bonsaglia, E. C. R., Silva, N. C. C., Fernandes Júnior, A., Araújo Júnior, J. P., Tsunemi, M. H. (2014). Production of biofilm by *Listeria monocytogenes* in different materials and temperatures. *Food Control*, 35, 386-391.

Borucki, M. K., Call, D. R. (2003). *Listeria monocytogenes* serotype identification by PCR. *Journal of Clinical Microbiology*, 41(12), 5537-5540.

Borucki, M. K., Peppin, J. D., White, D., Loge, F., Call, D. R. (2003). Variation in biofilm formation among strains of *Listeria monocytogenes*. *Applied and Environmental Microbiology*, 69, 7336-7342.

Bouayad, L., Hamdi, T.-M. (2012). Prevalence of *Listeria* spp. in ready to eat foods (RTE) from Algiers (Algeria). *Food Control*, 23(2), 397-399.

Boyle, K. E., Heilmann, S., van Ditmarsch, D., Xavier, J. B. (2013). Exploiting social evolution in biofilms. *Current Opinion in Microbiology*, 16, 207-212.

Bremer, P. J., Monk, I., Osborne, C. M. (2001). Survival of *Listeria monocytogenes* attached to stainless steel surfaces in the presence or absence of *Flavobacterium* spp. *Journal of Food Protection*, 64(9), 1369-1376.

Briandet, R., Leriche, V., Carpentier, B., Bellon-Fontaine, M. N. (1999a). Effects of the growth procedure on the surface hydrophobicity of *Listeria monocytogenes* cells and their adhesion to stainless steel. *Journal of Food Protection*, 62, 994-998.

Briandet, R., Meylheuc, T., Maher, C., Bellon-Fontaine, M. N. (1999b). *Listeria monocytogenes* Scott A: cell surface charge, hydrophobicity, and electron donor and acceptor characteristics under different environmental growth conditions. *Applied and Environmental Microbiology*, 65, 5328-5333.

Briandet, R., Lacroix-Gueu, P., Renault, M., Lecart, S., Meylheuc, T., Bidnenko, E., Steenkeste, K., Bellon-Fontaine, M. N., Fontaine-Aupart, M. P. (2008). Fluorescence correlation spectroscopy to study diffusion and reaction of bacteriophages inside biofilms. *Applied and Environmental Microbiology*, 74, 2135-2143.

Bridier, A., Sanchez-Vizuete, P., Guilbaud, M., Piard, J. C., Naïtali, M., Briandet, R. (2015). Biofilm-associated persistence of food-borne pathogens. *Food Microbiology*, 45, 167-178.

Burmolle, M., Webb, J. S., Rao, D., Hansen, L. H., Sorensen, S. J., Kjelleberg, S. (2006). Enhanced biofilm formation and increased resistance to antimicrobial agents and bacterial invasion are caused by synergistic interactions in multispecies biofilms. *Applied and Environmental Microbiology*, 72, 3916-3923.

Carpentier, B., Cerf, O. (2011). Review-Persistence of *Listeria monocytogenes* in food industry equipment and premises. *International Journal of Food Microbiology*, 145, 1-8.

Casarin, L. S., Brandelli, A., Casarin, F. de O., Soave, P. A., Wanke, C. H., Tondo, E. C. (2014). Adhesion of *Salmonella* Enteritidis and *Listeria monocytogenes* on stainless steel welds. *International Journal of Food Microbiology*, 191, 103-108.

CDC - Center for Disease Control and Prevention (2011). Vital signs: Incidence and trends of infection with pathogens transmitted commonly through food — Foodborne diseases active surveillance network, 10 US sites, 1996-2010. *Morbidity and Mortality Weekly Report*, 22, 749-755.

CDC - Center for Disease Control and Prevention (2014). National *Listeria* Surveillance annual summary 2012. Atlanta, Georgia: US Department of Health and Human Services, CDC. Available in: http://www.cdc.gov/listeria/pdf/listeria-annual-summary-2012-508c.pdf.

Cokal, Y., Dagdelen, A., Cenet, O., Gunsen, U. (2012). Presence of *L. monocytogenes* and some bacterial pathogens in two Turkish traditional foods, Mihalic cheese and Hosmerim dessert. *Food Control*, 26(2), 337-340.

Combrouse, T., Sadovskaya, I., Faille, C., Kol, O., Guérardel, Y., Midelet-Bourdin, G. (2013). Quantification of the extracellular matrix of the *Listeria monocytogenes* biofilms of different phylogenic lineages with optimization of culture conditions. *Journal of Applied Microbiology*, 114, 1120-1131.

Costerton, J. W., Lewandowski, Z., Caldwell, D. E., Korber, D. R., Lappin-Scott, H. M. (1995). Microbial biofilms. *Annual Review of Microbiology*, 49, 711-745.

Costerton, J. W., Lewandowski, Z., DeBeer, D., Caldwell, D., Korber, D., James, G. (1994). Biofilms, the customized microniche. *Journal of Bacteriology*, 176(8), 2137-2142.

Davey, M. E., O' Toole, G. A. (2000). Microbial biofilms: From ecology tomolecular genetics. *Microbiology and Molecular Biology Reviews*, 64(4), 847-867.

den Bakker, H. C., Warchocki, Wright, E. M., Allred, A. F., Ahlstrom, C., Manuel, C. S., Stasiewicz, M. J., Burrell, A., Roof, S., Strawn, L., Fortes, E. D., Nightingale, K. K.,

Kephart, D., Wiedmann, M. (2014). Five new species of *Listeria (L. floridensis* sp. nov., *L. aquatica* sp. nov., *L. cornellensis* sp. nov., *L. riparia* sp. nov., and *L. grandensis* sp. nov.) from agricultural and natural environments in the United States. *International Journal of Systematic and Evolutionary Microbiology*, 64, 1882-1889.

de Santis, E. P. L., Pilo, L., Cosseddu, a M., Canu, N., Scarano, C., Marongiu, P. (2007). Multiplex PCR for the identification and serotyping of *L. monocytogenes* isolated from sheep cheese-processing plants. *Veterinary Research Communications*, 1(31 Suppl.), 359-363.

Di Bonaventura, G., Piccolomini, R., Paludi, D., D'Orio, V., Vergara, A., Conter, M., Ianieri, A. (2008). Influence of temperature on biofilm formation by *Listeria monocytogenes* on various food-contact surfaces: Relationship with motility and cell surface hydrophobicity. *Journal of Applied Microbiology*, 104(6), 1552-1561.

Djordjevic, D., Wiedmann, M., McLandsborough, L. A. (2002). Microtiter plate assay for assessment of *Listeria monocytogenes* biofilm formation. *Applied and Environmental Microbiology*, 68(6), 2950-2958.

Donlan, R. M. (2002). Biofilms: microbial life on surfaces. *Emerging Infectious Diseases*, 8 (9), 881-890.

Donlan, R. M., Costerton, J. W. (2002). Biofilms: Survival mechanisms of clinically relevant microorganisms. *Clinical Microbiology Reviews*, 15(2), 167-193.

Donlan, R. M. (2009). Preventing biofilms of clinically relevant organisms using bacteriophage. *Trends in Microbiology*, 17, 66-72.

Doumith, M., Buchrieser, C., Glaser, P., Jacquet, C., Martin, P. (2004). Differentiation of the major *Listeria monocytogenes* serovars by multiplex PCR. *Journal of Clinical Microbiology*, 42(8), 3819-3822.

Duan, J., Park, S. I., Daeschel, M. A., Zhao, Y. (2007). Antimicrobial chitosan-lysozyme (CL) films and coatings for enhancing microbial safety of mozzarella cheese. *Journal of Food Science*, 72, M355-M362.

Dueholm, M. S., Petersen, S. V., Sonderkaer, M., Larsen, P., Christiansen, G., Hein, K. L., Enghild, J. J., Nielsen, J. L., Nielsen, K. L., Nielsen, P. H., Otzen, D. E. (2010). Functional amyloid in *Pseudomonas*. *Molecular Microbiology*, 77, 1009-1020.

Ebner, R., Stephan, R., Althaus, D., Brisse, S., Maury, M., Tasara, T. (2015) Phenotypic and genotypic characteristics of *Listeria monocytogenes* strains isolated during 2011 e 2014 from different food matrices in Switzerland. *Food Control*, 57, 321-326.

ECDC - European Centre for Disease Prevention and Control (2013). *Annual Epidemiological Report 2013. Reporting on 2011 surveillance data and 2012 epidemic intelligence data*. Stockholm: ECDC. Available in: http://ecdc.europa.eu/en/publications/ Publications/AnnualEpidemiologicalReport-2013.pdf.

EFSA and ECDC (European Food Safety Authority and European Centre for Disease Prevention and Control). (2014). The European Union summary report on trends and sources of Zoonoses, Zoonotic agents and food-borne outbreaks in 2012. *EFSA Journal 2015*, 13(1), 162. Available in: http://dx.doi.org/10.2903/j.efsa.2015.3991.

EFSA - European Food Safety Authority. (2015). The European Union summary report on trends and sources of zoonoses, zoonotic agents and food-borne outbreaks in 2013. *EFSA Journal*, 13(1):3991, 162 pp.

Eifert, J. D., Curtis, P. A., Bazaco, M. C., Meinersmann, R. J., Berrang, M. E., Kernodle, S., et al. (2005). Molecular characterization of *Listeria monocytogenes* of the serotype 4b complex (4b, 4d, 4e) from two turkey processing plants. *Foodborne and Pathogen Diseases*, 2, 192-200.

Ferreira, V., Barbosa, J., Stasiewicz, M., Vongkamjan, K., Moreno Switt, A., Hogg, T., et al. (2011). Diverse geno-and phenotypes of persistent *Listeria monocytogenes* isolates from fermented meat sausage production facilities in Portugal. *Applied and Environmental Microbiology*, 77, 2701-2715.

Flemming, H. C. (2011). The perfect slime. *Colloids and Surfaces B: Biointerfaces* 86, 251-259.

Flemming, H. C., Wingender, J. (2010). The biofilm matrix. *Nature Reviews Microbiology*, 8, 623-633.

Filiousis, G., Johansson, A., Frey, J., Perreten, V. (2009). Prevalence, genetic diversity and antimicrobial susceptibility of *Listeria monocytogenes* isolated from open-air food markets in Greece. *Food Control*, 20(3), 314-317.

Flint, S. H. (1996). Report on an evaluation of seven plasma weld samples. In: W. Scholz, *Stainless steel weld surface finish and biofilm development.* Around robin test, New Zealand Welding Center Report R8-14, (Section 5, pp. 1-7). Manukau City, New Zealand.

Garrido, V., Vitas, I., García-Jalon, I. (2009). Survey of *Listeria monocytogenes* in ready-to-eat products: prevalence by brands and retail establishments for exposure assessment of listeriosis in Northern Spain. *Food Control*, 20(11), 986-991.

Garsin, D. A., Willems, R. J. (2010). Insights into the biofilm lifestyle of enterococci. *Virulence*, 1, 219-221.

Giaouris, E., Chorianopoulos, N., Doulgeraki, A., Nychas, G. J. (2013). Co-Culture with *Listeria monocytogenes* within a dual-species biofilm community strongly increases resistance of *Pseudomonas putida* to benzalkonium chloride. *PloS One*, 8, e77276.

Giaouris, E., Heir, E., Hébraud, M., Chorianopoulos, N., Langsrud, S., Møretrø, T., Habimana, O., Desvaux, M., Renier, S., Nychas, G. (2014). Attachment and biofilm formation by foodborne bacteria in meat processing environments: Causes, implications, role of bacterial interactions and control by alternative novel methods. *Meat Science*, 97, 298-309.

Gibbons, I.-S., Adesiyun, A., Seepersadsingh, N., Rahaman, S. (2006). Investigation for possible source(s) of contamination of ready-to-eat meat products with *Listeria* spp. and other pathogens in a meat processing plant in Trinidad. *Food Microbiology*, 23(4), 359-366.

Gilbert, P., Allison, D. G., McBain, A. J. (2002). Biofilms in vitro and in vivo: Do singular mechanisms imply cross-resistance? *Journal of Applied Microbiology*, 92(s1), 98S-110S.

Guiton, P. S., Hung, C. S., Kline, K. A., Roth, R., Kau, A. L., Hayes, E., Heuser, J., Dodson, K. W., Caparon, M. G., Hultgren, S. J. (2009). Contribution of autolysin and sortase A during *Enterococcus faecalis* DNA-dependent biofilm development. *Infection and Immunity*, 77, 3626-3638.

Habimana, O., Meyrand, M., Meylheuc, T., Kulakauskas, S., Briandet, R. (2009). Genetic features of resident biofilms determine attachment of *Listeria monocytogenes*. *Applied and Environmental Microbiology*, 75, 7814-7821.

Habimana, O., Guillier, L., Kulakauskas, S., Briandet, R. (2011). Spatial competition with *Lactococcus lactis* in mixed-species continuous-flow biofilms inhibits *Listeria monocytogenes* growth. *Biofouling*, 27, 1065e 1072.

Hall-Stoodley, L., Stoodley, P. (2002). Developmental regulation of microbial biofilms. *Current Opinion in Biotechnology*, 13, 228-233.

Harakeh, S., Saleh, I., Zouhairi, O., Baydoun, E., Barbour, E., Alwan, N. (2009). Antimicrobial resistance of *Listeria monocytogenes* isolated from dairy-based food products. *Science of the Total Environment*, 407(13), 4022-4027.

Harmsen, M., Lappann, M., Knochel, S., Molin, S. (2010). Role of extracellular DNA during biofilm formation by *Listeria monocytogenes. Applied and Environmental Microbiology*, 76, 2271-2279.

Harvey, J., Keenan, K. P., Gilmour, A. (2007). Assessing biofilm formation by *Listeria monocytogenes* strains. *Food Microbiology*, 24, 380-392.

Hofer, E., Reis, C. M. F., Hofer, C. B. (2006). Serovars of *Listeria monocytogenes* and related species isolated from human clinical specimens. *Revista da Sociedade Brasileira de Medicina Tropical*, 39(1), 32-37.

Huang, B., Fang, N., Dimovski, K., Wang, X., Hogg, G., Bates, J. (2011). Observation of a new pattern in serogroup-related PCR typing of *Listeria monocytogenes* 4b isolates. *Journal of Clinical Microbiology*, 49(1), 426-429.

Izano, E. A., Amarante, M. A., Kher, W. B., Kaplan, J. B. (2008). Differential roles of poly-N-acetylglucosamine surface polysaccharide and extracellular DNA in *Staphylococcus aureus* and *Staphylococcus epidermidis* biofilms. *Applied and Environmental Microbiology*, 74, 470-476.

Jabbouri, S., Sadovskaya, I. (2010). Characteristics of the biofilm matrix and its role as a possible target for the detection and eradication of *Staphylococcus epidermidis* associated with medical implant infections. *FEMS Immunology and Medical Microbiology*, 59, 280-291.

Jahid, I. K., Ha, S. D. (2012). A review of microbial biofilms of produce: future challenge to food safety. *Food Science and Biotechnology*, 21, 299-316.

Jakobsen, R. A., Heggebø, R., Sunde, E. B., Skjervheim, M. (2011). *Staphylococcus aureus* and *Listeria monocytogenes* in Norwegian raw milk cheese production. *Food Microbiology*, 28(3), 492-496.

Jamali, H., Radmehr, B., Thong, K. L. (2013). Prevalence, characterization, and antimicrobial resistance of *Listeria* species and *Listeria monocytogenes* isolates from raw milk in farm bulk tanks. *Food Control*, 34(1), 121-125.

Jensen, A., Larsen, M. H., Ingmer, H., Vogel, B. F., Lone, G. (2007). Sodium chloride enhances adherence and aggregation and strain variation influences invasiveness of *Listeria monocytogenes* strains. *Journal of Food Protection*, 70, 592-599.

Kadam, S. R., den Besten, H. M. W., van der Veen, S., Zwietering, M. H., Moezelaar, R., Abee, T. (2013). Diversity assessment of *Listeria monocytogenes* biofilm formation: Impact of growth condition, serotype and strain origin. *International Journal of Food Microbiology*, 165, 259-264.

Kathariou, S. (2002). *Listeria monocytogenes* virulence and pathogenicity, a food safety perspective. *Journal of Food Protection*, 65(11), 1811-1829.

Korea, C. G., Ghigo, J. M., Beloin, C. (2011). The sweet connection: solving the riddle of multiple sugar-binding fimbrial adhesins in *Escherichia coli*: multiple *E. coli* fimbriae form a versatile arsenal of sugar-binding lectins potentially involved in surface-colonisation and tissue tropism. *BioEssays: News and Reviews in Molecular, Cellular and Developmental Biology*, 33, 300-311.

Kim, T. J., Young, B. M., Young, G. M. (2008). Effect of flagellar mutations on *Yersinia enterocolitica* biofilm formation. *Applied and Environmental Microbiology*, 74(17), 5466-5574.

Kumar, C. G., Anand, S. K. (1998). Significance of microbial biofilms in food industry: a review. *International Journal of Food Microbiology*, 42(1-2), 9-27.

Lang Halter, E., Neuhaus, K., Scherer, S. (2013) *Listeria weihenstephanensis* sp. nov., isolated from the water plant *Lemna trisulca* taken from a freshwater pond. *International Journal of Systematic and Evolutionary Microbiology*, 63, 641-647.

Lappi, V. R., Thimothe, J., Nightingale, K. K., Gall, K., Scott, V. N., Wiedmann, M. (2004). Longitudinal studies on *Listeria* in smoked fish plants: impact of intervention strategies on contamination patterns. *Journal of Food Protection*, 67, 2500-2514.

Leclercq, A., Clermont, D., Bizet, C., Grimont, P. A. D., Le Flèche-Matéos, A., Roche, S. M., Buchrieser, C., Cadet-Daniel, V., LeMonnier, A., Lecuit, M., Allerberger, F. (2010). *Listeria rocourtiae* sp. nov. *International Journal of Systematic and Evolutionary Microbiology*, 60, 2210-2214.

Lee, S., Ahn, J., Lee, K., Um, H., Sekhon, S. S., Park, T. S., Min, J., Kim, Y. (2015). Analytical bioconjugates, aptamers, enable specific quantitative detection of *Listeria monocytogenes. Biosensors and Bioelectronics*, 68, 272-280.

Lemon, K. P., Higgins, D. E., Kolter, R. (2007). Flagellar motility is critical for *Listeria monocytogenes* biofilm formation. *Journal of Bacteriology*, 189, 4418-4424.

Lequette, Y., Boels, G., Clarisse, M., Faille, C. (2010). Using enzymes to remove biofilms of bacterial isolates sampled in the food-industry. *Biofouling*, 26, 421-431.

Leriche, V., Chassaing, D., Carpentier, B. (1999). Behaviour of *L. monocytogenes* in an artificially made biofilm of a nisin-producing strain of Lactococcus lactis. *International Journal of Food Microbiology*, 51, 169-182.

Liu, D. (2006). Identification, subtyping and virulence determination of *Listeria monocytogenes*, an important foodborne pathogen. *Journal of Medical Microbiology*, 55 (6), 645-659.

Lu, T. K., Collins, J. J. (2007). Dispersing biofilms with engineered enzymatic bacteriophage. *Proceedings of the National Academy of Sciences US*, 104, 11197-11202.

Lundén, J., Autio, T., Markkula, A., Hellström, S., Korkeala, H. (2003). Adaptive and cross-adaptive responses of persistent and non-persistent *Listeria monocytogenes* strains to disinfectants. *International Journal of Food Microbiology*, 82, 265-272.

Luppens, S. B. I., Kara, D., Bandounas, L., Jonker, M. J., Wittink, F. R. A., Bruning, O., Breit, T. M., ten Cate, J. M., Crielaard, W. (2008). Effect of *Veillonella parvula* on the antimicrobial resistance and gene expression of *Streptococcus mutans* grown in a dual-species biofilm. *Oral Microbiol. Immunology*, 23, 183-189.

Mackiw, E., Modzelewska, M., Maka, L., Sciezynska, H., Pawłowska, K., Postupolski, J., Korsak, D. (2016) Antimicrobial resistance profiles of Listeria monocytogenes isolated from ready-to-eat products in Poland in 2007-2011. *Food Control*, 59, 7-11.

Mandlik, A., Swierczynski, A., Das, A., Ton-That, H. (2008). Pili in Gram-positive bacteria: assembly, involvement in colonization and biofilm development. *Trends in Microbiology*, 16, 33-40.

Mariani, C., Briandet, R., Chamba, J. F., Notz, E., Carnet-Pantiez, A., Eyoug, R. N., Oulahal, N. (2007). Biofilm ecology of wooden shelves used in ripening the French raw milk smear cheese Reblochon de Savoie. *Journal of Dairy Science*, 90, 1653-1661.

Marshall, K. C. (1992). Biofilms: and overview of bacterial adhesion, activity, and control at surfaces. *American Society for Microbiology*, 58, 202-207.

Martins, E. A., Leal Germano, P. M. (2011). *Listeria monocytogenes* in ready-to-eat, sliced, cooked ham and salami products, marketed in the city of São Paulo, Brazil: occurrence, quantification, and serotyping. *Food Control*, 22(2), 297-302.

Mazmanian, S. K., Liu, G., Ton-That, H., Schneewind, O. (1999). *Staphylococcus aureus* sortase, an enzyme that anchors surface proteins to the cell wall. *Science*, 285, 760-763.

Melo, J., Andrewb, P. W., Faleiro, M. L. (2015). *Listeria monocytogenes* in cheese and the dairy environment remains a food safety challenge: The role of stress responses. *Food Research International*, 67, 75-90.

Meyer, B. (2003). Approaches to prevention, removal and killing of biofilms. *International Biodeterioration and Biodegradation*, 51(4), 249-253.

Meylheuc, T., Methivier, C., Renault, M., Herry, J. M., Pradier, C. M., Bellon- Fontaine, M. N. (2006). Adsorption on stainless steel surfaces of biosurfactants produced by gram-negative and gram-positive bacteria: consequence on the bioadhesive behavior of *Listeria monocytogenes*. *Colloids and Surfaces B: Biointerfaces*, 52, 128-137.

Midelet, G., Carpentier, B. (2004). Impact of cleaning and disinfection agents on biofilm structure and on microbial transfer to a solid model food. *Journal of Applied Microbiology*, 97(2), 262-270.

Moens, S., Vanderleyden, J. (1996). Functions of bacterial flagella. *Critical Reviews in Microbiology*, 22(2), 67-100.

Molin, S., Tolker-Nielsen, T. (2003). Gene transfer occurs with enhanced efficiency in biofilms and induces enhanced stabilisation of the biofilm structure. *Current Opinion in Biotechnology*, 14(3), 255-261.

Neyret, C., Herry, J. M., Meylheuc, T., Dubois-Brissonnet, F. (2014). Plant-derived compounds as natural antimicrobials to control paper mill biofilms. *Journal of Industrial Microbiology and Biotechnology*, 41, 87-96.

Ni, N., Li, M., Wang, J., Wang, B. (2009). Inhibitors and antagonists of bacterial quorum sensing. *Medicinal Research Reviews*, 29, 65-124.

Nilsson, R. E., Ross, T., Bowman, J. P. (2011). Variability in biofilm production by *Listeria monocytogenes* correlated to strain origin and growth conditions. *International Journal of Food Microbiology*, 150, 14-24.

Orgaz, B., Neufeld, R. J., SanJose, C. (2007). Single-step biofilm removal with delayed release encapsulated pronase mixed with soluble enzymes. *Enzyme and Microbial Technology*, 40, 1045-1051.

Pan, Y., Breidt, F., Gorski, L. (2010). Synergistic effects of sodium chloride, glucose, and temperature on biofilm formation by *Listeria monocytogenes* serotype 1/2a and 4b strains. *Applied and Environmental Microbiology,* 76, 1433-1441.

Peel, M., Donachie, W., Shaw, A. (1988). Temperature-dependent expression of flagella of *Listeria monocytogenes* studied by electron microscopy, SDS-PAGE and western blotting. *Journal of General Microbiology*, 134(8), 2171-2178.

Pilchová, T., Hernould, M., Prévost, H., Demnerová, K., Pazlarová, J., Tresse, O. (2014). Influence of food processing environments on structure initiation of static biofilm of *Listeria monocytogenes*. *Food Control*, 35 366-372.

Poncin-Epaillard, F., Herry, J. M., Marmey, P., Legeay, G., Debarnot, D., Bellon-Fontaine, M. N. (2013). Elaboration of highly hydrophobic polymeric surface—a potential strategy to reduce the adhesion of pathogenic bacteria? *Materials Science and Engineering C*, 33, 152-1161.

Pratt, L. A., Kolter, R. (1998). Genetic analysis of *Escherichia coli* biofilm formation: roles of flagella, motility, chemotaxis and type I pili. *Molecular Microbiology*, 30, 285-293.

Rahimi, E., Ameri, M., Momtaz, H. (2010). Prevalence and antimicrobial resistance of *Listeria* species isolated from milk and dairy products in Iran. *Food Control*, 21(11), 1448-1452.

Renier, S., Hébraud, M., Desvaux, M. (2011). Molecular biology of surface colonization by *Listeria monocytogenes*: an additional facet of an opportunistic Gram-positive foodborne pathogen. *Environmental Microbiology*, 13(4), 835-850.

Renier, S., Chagnot, C., Deschamps, J., Caccia, N., Szlavik, J., Joyce, S. A., Popowska, M., Hill, C., Knochel, S., Briandet, R., Hebraud, M., Desvaux, M. (2013). Inactivation of the SecA2 protein export pathway in *Listeria monocytogenes* promotes cell aggregation, impacts biofilm architecture and induces biofilm formation in environmental condition. *Environmental Microbiology*, 16, 1176-1192.

Rieu, A., Weidmann, S., Garmyn, D., Piveteau, P., Guzzo, J. (2007). Agr system of *Listeria monocytogenes* EGD-e: role in adherence and differential expression pattern. *Applied and Environmental Microbiology,* 73, 6125-6133.

Rieu, A., Briandet, R., Habimana, O., Garmyn, D., Guzzo, J., Piveteau, P. (2008a). *Listeria monocytogenes* EGD-e biofilms: no mushrooms but a network of knitted chains. *Applied and Environmental Microbiology*, 74(14), 4491-4497.

Rieu, A., Lemaitre, J. P., Guzzo, J., Piveteau, P. (2008b). Interactions in dual species biofilms between *Listeria monocytogenes* EGD-e and several strains of Staphylococcus aureus. *International Journal of Food Microbiology*, 126, 76-82.

Romero, D., Aguilar, C., Losick, R., Kolter, R. (2010). Amyloid fibers provide structural integrity to *Bacillus subtilis* biofilms. *Proceedings of the National Academy of Sciences US*, 107, 2230-2234.

Sachs, J. L., Hollowell, A. C. (2012). The origins of cooperative bacterial communities. *MBio*, 3, e00099-12.

Sauer, K., Rickard, A. H., Davies, D. G. (2007). Biofilms and biocomplexity. *American Society for Microbiology*, 2, 347-355.

Schwering, M., Song, J., Louie, M., Turner, R. J., Ceri, H. (2013). Multi-species biofilms defined from drinking water microorganisms provide increased protection against chlorine disinfection. *Biofouling*, 29, 917-928.

Setiani, B. E., Elegado, F. B., Perez, M. T. M., Mabesa, R. C., Dizon, E. I., Sevilla, C. C. (2015) API Listeria Rapid kit for Confirmatory Fenotypic Conventional Biochemical Test of the Prevalence *Listeria monocytogenes* in selected meat and meat products. *Procedia Food Science*, 3, 445-452.

Sillanpaa, J., Nallapareddy, S. R., Singh, K. V., Prakash, V. P., Fothergill, T., Ton-That, H., Murray, B. E. (2010). Characterization of the ebp(fm) pilus-encoding operon of *Enterococcus faecium* and its role in biofilm formation and virulence in a murine model of urinary tract infection. *Virulence*, 1, 236-246.

Silva, S., Teixeira, P., Oliveira, R., Azeredo, J. (2008). Adhesion to and viability of *Listeria monocytogenes* on food contact surfaces. *Journal of Food Protection*, 71, 1379-1385.

Simoes, M., Simoes, L. C., Machado, I., Pereira, M. O., Vieira, M. J. (2006). Control of flow-generated biofilms with surfactants: evidence of resistance and recovery. *Food and Bioproducts*, 84(4), 338-345.

Simoes, M., Simoes, L. C., Vieira, M. J. (2010). A review of current and emergent biofilm control strategies. *LWT- Food Science and Technology*, 43, 573-583.

Srey, S., Jahid, I. K., Ha, S. (2013). Biofilm formation in food industries: A food safety concern. *Food Control*, 31, 572-585.

Szlavik, J., Paiva, D. S., Mørk, N., van den Berg, F., Verran, J., Whitehead, K., Knøchel, S., Niels, D. S. (2012). Initial adhesion of *Listeria monocytogenes* to solid surfaces under liquid flow. *International Journal of Food Microbiology*, 152, 181-188.

Takahashi, H., Suda, T., Tanaka, Y., Kimura, B. (2010). Cellular hydrophobicity of *Listeria monocytogenes* involves initial attachment and biofilm formation on the surface of polyvinyl chloride. *Letters in Applied Microbiology*, 50(6), 618-625.

Takhistov, P., George, B. (2004). Linearized kinetic model of *Listeria monocytogenes* biofilm growth. *Bioprocess and Biosystems Engineering*, 26(4), 259-270.

Tide, C., Harkin, S. R., Geesey, G. G., Bremer, F. J., Scholz, W. (1999). The influence of welding procedures on bacterial colonization of stainless steel weldments. *Journal of Food Engineering*, 42, 85-96.

Todhanakasem, T., Young, G. M. (2008). Loss of flagellum-based motility by *Listeria monocytogenes* results in formation of hyperbiofilms. *Journal of Bacteriology*, 190, 6030-6034.

Tompkin, R. B. (2002). Control of *Listeria monocytogenes* in the food-processing environment. *Journal of Food Protection*, 65, 709-725.

Tresse, O., Shannon, K., Pinon, A., Malle, P., Vialette, M., Midelet-Bourdin, G. (2007). Variable adhesion of *Listeria monocytogenes* isolates from food processing facilities and clinical cases to inert surfaces. *Journal of Food Protection*, 70(7), 1569-1578.

Uhitil, S., Jaksic, S., Petrak, T., Medic, H., Gumhalter-Karolyi, L. (2004). Prevalence of *Listeria monocytogenes* and the other *Listeria* spp. in cakes in Croatia. *Food Control*, 15 (3), 213-216.

van der Veen, S., Abee, T. (2011). Mixed species biofilms of *Listeria monocytogenes* and *Lactobacillus plantarum* show enhanced resistance to benzalkonium chloride and peracetic acid. *International Journal of Food Microbiology*, 144, 421-431.

van Houdt, R., Michiels, C. W. (2005). Role of bacterial cell surface structures in *Escherichia coli* biofilm formation. *Research in Microbiology*, 156(5-6), 626-633.

van Houdt, R., Michiels, C. W. (2010). Biofilm formation and the food industry, a focus on the bacterial outer surface. *Journal of Applied Microbiology*, 109, 1117-1131.

van Loosdrecht, M. C. M., Lyklema, J., Norde, W., Zehnder, A. J. B. (1989). Bacterial adhesion: a physicochemical approach. *Microbial Ecology*, 17, 1-15.

Vatanyoopaisarn, S., Nazli, A., Dodd, C. E., Rees, C. E., Waites, W. M. (2000). Effect of flagella on initial attachment of *Listeria monocytogenes* to stainless steel. *Applied and Environmental Microbiology*, 66(2), 860-863.

Vazquez-Boland, J. A., Kuhn, M., Berche, P., Chakraborty, T., Domínguez-Bernal, G., Goebel, W., et al. (2001). *Listeria* pathogenesis and molecular virulence determinants. *Clinical Microbiology Reviews*, 14(3), 584-640.

Vilain, S., Pretorius, J. M., Theron, J., Brozel, V. S. (2009). DNA as an adhesin: *Bacillus cereus* requires extracellular DNA to form biofilms. *Applied and Environmental Microbiology*, 75, 2861-2868.

Vogel, B. F., Hansen, L. T., Mordhorst, H., Gram, L. (2010). The survival of *Listeria monocytogenes* during long term desiccation is facilitated by sodium chloride and organic material. *International Journal of Food Microbiology*, 140, 192-200.

Watnick, P., Kolter, R. (2000). *Biofilm, city of microbes. Journal of Bacteriology*, 182(10), 2675-2679.

Weller, D., Andrus, A., Wiedmann, M., den Bakker, H. (2015) *Listeria booriae* sp. nov., and *Listeria newyorkensis* sp. nov., from food processing environments in the United States. *International Journal of Systematic and Evolutionary Microbiology*, 65, 286-292.

Williams, S. K., Roof, S., Boyle, E. A., Burson, D., Thippareddi, H., Geornaras, I., et al. (2011). Molecular ecology of *Listeria monocytogenes* and other *Listeria* species in small and very small ready-to-eat meat processing plants. *Journal of Food Protection*, 74, 63-77.

Winkelströter, L. K., Gomes, B. C., Thomaz, M. R. S., Souza, V. M., De Martinis, E. C. P. (2011). *Lactobacillus sakei* 1 and its bacteriocin influence adhesion of *Listeria monocytogenes* on stainless steel surface. *Food control*, 22, 1404-1407.

Yarwood, J. M., Bartels, D. J., Volper, E. M., Greenberg, E. P. (2004). Quorum sensing in *Staphylococcus aureus* biofilms. *Journal Bacteriology*, 186, 1838-1850.

Xavier, J. B., Picioreanu, C., Rani, S. A., van Loosdrecht, M. C., Stewart, P. S. (2005). Biofilm-control strategies based on enzymic disruption of the extracellular polymeric substance matrixea modelling study. *Microbiology*, 151, 3817-3832.

Xavier, J. B., Foster, K. R. (2007). Cooperation and conflict in microbial biofilms. *Proceedings of the National Academy of Sciences US*, 104, 876-881.

Zhang, W., Chu, P. K., Ji, J., Zhang, Y., Fu, R. K. Y., Yan, Q. (2006). Antibacterial properties of plasma-modified and triclosan or bronopol coated polyethylene. *Polymer*, 47, 931-936.

In: Listeria monocytogenes
Editor: Tiffany Vicario

ISBN: 978-1-63483-765-1

Chapter 7

THE GASTROINTESTINAL PHASE OF *LISTERIA MONOCYTOGENES* INFECTION

Roy D. Sleator *
Department of Biological Sciences, Cork Institute of Technology,
Bishopstown, Cork, Ireland

ABSTRACT

Listeria monocytogenes is a ubiquitous bacterium that causes significant foodborne disease with high mortality rates in immunecompromised adults. Recent application of transcriptomic and proteomic approaches, as well as the development of new model systems, has enabled a greater focus on factors that influence adaptation to gastrointestinal environments and adhesion to and invasion of the gastrointestinal mucosa. Here I review some of the most exciting developments in the analysis of the interaction between *L. monocytogenes* and the host GI tract.

INTRODUCTION

Listeria monocytogenes is a Gram positive foodborne pathogen that is the causative agent of listeriosis, a disease which manifests primarily as meningitis or meningoencephalitis in non-pregnant individuals or infection of the foetus in pregnant women. *L. monocytogenes* normally resides in decaying plant matter in the soil but may be transiently associated with the GI tract in a number of animal species. Infection by, or indeed carriage of, the pathogen by domestic animals poses a risk of zoonotic transmission through contaminated milk or meat (Swaminathan and Gerner-Smidt 2007).

The bacterium encounters the mammalian GI tract at a number of stages during the infectious cycle; potentially during asymptomatic intestinal carriage in animals and subsequently during infection of the human host before invasion of the gastrointestinal

* Corresponding author: Department of Biological Sciences, Cork Institute of Technology, Bishopstown, Cork, Ireland. Phone: + 353-21-4335405 email: roy.sleator@cit.ie.

epithelium. The listerial genome reflects this ability to adapt to a variety of environments and contains a complex array of genes encoding regulatory proteins and virulence associated factors linked to survival in the GI tract, including bile salt hydrolase (*bsh*) and bile exclusion protein (*bilE*) (Begley et al. 2005; Dussurget et al. 2002; Glaser et al. 2001; Sleator et al. 2005). The current review analyses the most recent contributions to our understanding of the gastrointestinal phase of the listerial infectious process.

TRANSCRIPTOMIC ANALYSES AND INSIGHTS INTO ADAPTATION TO THE GI TRACT

The listerial *sigB* gene was independently isolated by Wiedmann et al. (1998) and Becker et al. (1998) on the bases of sequence homology with the *B. subtilis sigB* locus. While the former study identified a role for the encoded σ^B-protein in acid tolerance of *L. monocytogenes* 689426 (Wiedmann et al. 1998), the latter demonstrated that the activity of σ^B in *L. monocytogenes* Scott A is responsive to osmotic up-shift, temperature down-shift, and the presence of EDTA in the growth medium (Becker et al. 1998). The magnitude of the response was greatest after an osmotic up-shift, suggesting a role for σ^B in coordinating osmotic responses in *L. monocytogenes*. This observation was later confirmed by a partial gene array study comprising of 208-genes; consisting of 166 genes downstream of Hidden Markov Model (HMM) predicted σ^B -dependent promoters as well as selected virulence and stress response genes from *L. monocytogenes* 10403S (Kazmierczak et al. 2003). 55 genes with statistically significant σ^B-dependent expression were identified under stationary phase or conditions of osmotic stress, with at least 1.5-fold-higher expression in the wild type over the *sigB* mutant under either stress condition (Kazmierczak et al. 2003). A more recent whole genome array analysis of the σ^B regulon in *L. monocytogenes* EGD-e revealed 105 σ^B-positively regulated genes and 111 genes which appeared to be under negative control of σ^B at various stages of the growth cycle (Hain et al. 2008). The σ^B regulon therefore includes 7.6% of genes in the *L. monocytogenes* genome. Of the genes positively regulated by σ^B, 75 have homologues in *B. subtilis*, but only 33 had been previously described as σ^B-regulated in *B. subtilis*. The data indicate a divergence of the σ^B regulons between these two bacterial genera, possibly reflecting adaptation to differing niches (Hain et al. 2008).

In *Listeria*, the function of σ^B appears to be both species and strain dependent. While σ^B contributes to both stationary and log phase acid resistance in *L. monocytogenes*, only log phase acid resistance is σ^B–dependent in *L. innocua* (Raengpradub et al. 2008). Furthermore, compelling evidence indicates that the *sigB* gene plays a variable role in stress response in the different genetic lineages of *L. monocytogenes* (Moorhead & Dykes, 2003). It was shown that a serotype 1/2a strain is more reliant upon an intact σ^B regulon than a serotype 4c strain across a range of environmental stresses (Moorhead and Dykes 2003). Such variations in environmental stress resistance across different strains may contribute to disparate survival capabilities in foods and during infection and therefore indirectly to discrepancies in virulence potential between different strains (Wiedmann et al. 1998).

It has been shown that the σ^B regulon includes genes encoding the classical virulence factors InlA and InlB (Hain et al. 2008; Kazmierczak et al. 2003), and σ^B is required for rapid induction of expression of *L. monocytogenes* genes most likely important for survival of gastrointestinal stresses, including reduced pH, elevated osmolarity and bile salts. Indeed, loss of σ^B has been found to result in decreased virulence of *L. monocytogenes* following oral infection in the murine and guinea pig models but not during systemic infection (Garner et al. 2006; Nadon et al. 2002). It has been shown that the gene encoding σ^B is transcriptionally upregulated during transit through the mouse GI tract (Begley et al. 2005). Furthermore, the σ^B regulon is induced by many of the stresses encountered during gastrointestinal passage (Sue et al. 2004) and activates a number of systems that are important for bacterial homeostasis in the gut (Hain et al. 2008). σ^B is therefore critical for optimal listerial pathogenesis during gastrointestinal infection. Furthermore, this alternative sigma factor has the potential to promote bacterial survival, both inside and outside the host, thus contributing to survival at all stages of the infectious cycle (Gahan and Hill 2005; Sleator et al. 2003a).

Acid Tolerance

It appears that σ^B contributes to acid resistance through at least 2 mechanisms: firstly a general acid tolerance to which σ^B–regulated systems contribute throughout all growth phases, as well as a pH-inducible ATR mechanism that is at least partially σ^B-dependent in exponential phase cells (Ferreira et al. 2001; Volker et al. 1999). Becker et al. (1998) discerned from primer extension analyses that *sigB* expression in log phase *L. monocytogenes* can be induced from undetectable levels to a level similar to that observed for stationary phase cells following exposure of cells to mildly acidified media (pH 5.3). The glutamate decarboxylase system (GAD) plays an essential role in pH homeostasis in *L. monocytogenes* mediated primarily through a σ^B-regulated component GadD2 (formerly GadB) (Cotter et al. 2001; Kazmierczak et al. 2003; Wemekamp-Kamphuis et al. 2004). In addition, it has been shown that the arginine deiminase (ADI) acid tolerance system in *L. monocytogenes* is functional and is regulated by both σ^B and a dedicated transcriptional regulator, ArgR (Ryan et al. 2008; Ryan et al. 2009). The influence of these acid tolerance mechanisms upon *in vivo* gastric transit remains to be established; however it is likely that the glutamate decarboxylase (GAD) and ADI systems play a significant role in this process.

Bile Tolerance

The genes encoding the principal listerial bile resistance mechanisms, BSH and BilE, are preceded by σ^B dependent promoter sites and are transcriptionally down regulated in a *sigB* mutant (Begley et al. 2005; Sleator et al. 2005). Furthermore, BSH activity is completely abolished in a *L. monocytogenes* Δ*sigB* strain (Begley et al. 2005); thus suggesting a central role for σ^B in coordinating the listerial bile stress response that is necessary for small intestinal survival of the pathogen (Begley et al. 2005; Dussurget et al. 2002; Sleator et al. 2005).

Osmotolerance

The existence of putative σ^B-dependent promoter sites upstream of the osmolyte transporter genes *betL*, *gbu* and *opuC*, suggests that at least a component of osmolyte uptake in *L. monocytogenes* forms part of the σ^B regulon (Sleator et al. 2003b). Indeed, kinetic analysis of transcript accumulation after osmotic up-shift demonstrated that σ^B-dependent transcripts from *gbuA*P2 accumulate for an extended period after up-shift, suggesting that σ^B activity may provide a mechanism for sustained high-level expression during osmotic challenge (Cetin et al. 2004; Fraser et al. 2003). Furthermore, Becker et al. (1998, 2000) demonstrated that a σ^B mutant of *L. monocytogenes* is significantly impaired in its ability to use betaine and carnitine as osmoprotectants (Becker et al. 1998; Becker et al. 2000). I have previously shown that the principal carnitine transporter in *L. monocytogenes*, OpuC, is required for full virulence by the oral route (Sleator et al. 2001; Wemekamp-Kamphuis et al. 2002). Recent work has demonstrated that the *in vivo* requirement for this osmolyte uptake system is mediated through a role for OpuC in bile tolerance in the small intestine during oral infection in mice (Watson et al. 2009).

Cross Adaptation during Gastrointestinal Transit

Following ingestion, the first physical stress encountered by the bacterium is the low pH of the stomach (~pH2), followed by elevated osmolarity (equivalent to 0.3 M NaCl) and activity of the biological detergent bile in the upper small intestine. Significantly, pre-exposure to elevated osmolarity (0.3 M NaCl for 1 hour) resulted in a dramatic increase in the ability of *L. monocytogenes* to deal with lethal concentrations of bile (adapted cells surviving 1000 times better than naïve cells) (Begley et al. 2002; Sleator et al. 2007). However, a low pH challenge (as experienced during gastric transit) fails to similarly protect against subsequent osmotic or bile stress. Similarly, pre-exposure to bile fails to protect against acid or salt, but does protect against re-challenge with higher levels of bile salts (Begley et al. 2002). Thus, osmotic stress appears to be at the top of the hierarchy of stress responses during GI transit. Given that *sigB* is transcriptionally up-regulated at elevated osmolarity (Becker et al. 1998), it is likely that the increased osmolarity of the gastrointestinal lumen may be interpreted by *L. monocytogenes* as an environmental cue, signaling gut entry (as is the case for *Salmonella*) (Sleator and Hill 2002). Furthermore, given that σ^B has been shown to modulate expression of PrfA (Positive regulatory factor A; the master regulator of the virulence gene cluster which coordinates the intracellular phase of *L. monocytogenes* infection (Vazquez-Boland et al. 2001)) it is possible that osmotically induced stimulation of the σ^B regulon in the upper small intestine may not only facilitate successful gastrointestinal transit, but may also prime the pathogen for the next phase of infection which, in susceptible individuals, is the systemic invasive disease listeriosis (Ollinger et al. 2009).

σ^B Interplay with PrfA and Other Regulators

L. monocytogenes has the ability to adapt rapidly to a variety of environments, switching from a saprophyte in soil and decaying vegetation, to an intracellular bacterial pathogen capable of causing serious infection in humans and several animal species. This transformation is mediated by a complex interplay between the regulatory networks which modulate the expression of stress and virulence associated factors in response to specific environmental cues (Gahan and Hill 2005; Gray et al. 2006).

In this model σ^B dominates both in the external environment and during gastrointestinal transit, while it is PrfA which plays the central role during the next phase: intracellular infection. In addition to regulating the expression of stress response genes, σ^B (in concert with PrfA) also activates the transcription of virulence genes, such as *inlAB*, *bsh* and *bilE* (Begley et al. 2005; Kazmierczak et al. 2003; Sleator et al. 2005). Furthermore, σ^B directly regulates the transcription of *prfA* through a specific promoter ($P2_{prfA}$ promoter) (Nadon et al. 2002; Rauch et al. 2005; Schwab et al. 2005). There appears to be a dual role for σ^B as both an initial activator of *prfA* transcription (most likely during the gastrointestinal phase of infection) and a repressor of the PrfA regulon during intracellular infection. SigmaB is proposed as a means of fine-tuning expression of PrfA-regulated genes during cellular infection in order to reduce cytotoxicity in *Listeria*-infected cells (Ollinger et al. 2009).

In addition to modulating the activity of PrfA, σ^B has also been shown to form a regulatory network with the negative regulators HrcA and CtsR (class three stress gene repressor), involved in the regulation of heat shock genes. Microarray transcriptomic analyses and promoter searches identified at least 40 genes co-regulated by both CtsR and σ^B, including genes encoding proteins with confirmed or likely roles in virulence and stress response. These data demonstrate that interactions between CtsR and σ^B play an important role in *L. monocytogenes* stress resistance and virulence (Hu et al. 2007b). A similar approach revealed 31 genes co-regulated by HrcA and σ^B (Hu et al. 2007a). Thus, σ^B, CtsR and HrcA appear to form a regulatory network that contributes to the transcription of a number of *L. monocytogenes* genes.

Adherence to and Invasion of the Gastrointestinal Epithelium

L. monocytogenes has been demonstrated to bind to a specific human mucin, Muc2, but not to Muc1 (Linden et al. 2008). The pathogen interacts with Muc2 through internalin proteins with a requirement for these proteins of the order, InlB > InlC > InlJ (Linden et al. 2008). This initial interaction may be important for localizing the pathogen in the mucin layer prior to cellular infection. Significantly, InlJ also appears to act directly as a cell adhesin and heterologous expression of this protein in *L. innocua* promotes significant adherence to HT-29 and JEG-3 cells in vitro (Sabet et al. 2008). Expression of a 104 kDa protein called *Listeria* adhesion protein (LAP) is also important for adhesion of the pathogen to the host cell via interaction with host heat shock protein 60 (Wampler et al. 2004). Expression of LAP in *E. coli* conferred an enhanced ability to bind both Caco-2 cells and purified Hsp60 (Kim et al. 2006). Furthermore, the pathogen has the capacity to bind to cell surface fibronectin through a specific fibronectin binding protein (FbpA) which is important for murine infection via the oral route (Dramsi et al. 2004).

Localization of surface-associated virulence factors in *L. monocytogenes* is likely to be influenced by Auto, a cell wall hydrolase that is important for remodeling the listerial cell wall and is essential for cell invasion (Bublitz et al. 2009). *L. monocytogenes* expression of the sortase-anchored surface protein Vip is essential for infection of certain mammalian cell lines and for infection of mice by the oral route. The host receptor for Vip has been established to be the endoplasmic reticulum (ER) resident chaperone Gp96 (Cabanes et al. 2005). It has also been well established that *L. monocytogenes* requires the surface protein Internalin A in order to invade gastrointestinal enterocytes. Internalin B does not appear to play a role in invasion of enterocytes but is required for invasion of hepatocytes and infection of the fetoplacental unit (Disson et al. 2008; Khelef et al. 2006). It has long been established that the host receptor for InlA is E-cadherin, a protein expressed towards the basolateral surface of polarized enterocytes (Mengaud et al. 1996). Further work has demonstrated that InlA has the opportunity to interact with E-cadherin during extrusion of apoptotic cells at the villous tips, a process which transiently makes E-cadherin available for binding by L. monocytogenes expressing InlA (Pentecost et al. 2006). The molecular mechanisms underpinning InlA-mediated cell entry have been elucidated and extensively reviewed elsewhere (Bonazzi et al. 2009; Hamon et al. 2006; Ireton, 2007; Seveau et al. 2007).

Role of the Microbiota in Resistance to Infection

The human gastrointestinal tract contains a vast array of microorganisms (between 500-1000 species) which reach highest levels in the colon (approximately 10^{11}-10^{12} microorganisms per gram of contents). The small intestine (the site of listerial invasion) also contains a sizeable microbiota with between 10^4 and 10^8 microorganisms per gram (Rastall 2004). The gut microbiota contributes to nutrient acquisition, development of the immune system and plays a significant role in resistance to foodborne pathogens (Xu and Gordon 2003). The presence of a gut microbiota contributes to protection against orally administered *L. monocytogenes* in rodent models of disease (Czuprynski and Balish 1981; Vieira et al. 2008). Furthermore, immune responses to *L. monocytogenes* are blunted in mice raised under gnotobiotic conditions, and these mice demonstrate elevated susceptibility to systemic (ip) infection (Inagaki et al. 1996).

The concept of using probiotic commensals to protect against oral infection with *L. monocytogenes* has been investigated using a number of model systems. A variety of potentially probiotic strains including *E. coli* Nissle 1917 (Altenhoefer et al. 2004) *Lactobacillus salivarius* UCC118 and *Bifidobacterium breve 2003* (Corr et al. 2007a) have been shown to inhibit *L. monocytogenes* invasion of gastrointestinal cell culture lines. Oral administration of live *L. delbrueckii* UFV-H2b20 has been shown to significantly protect against *L. monocytogenes* infection in a mouse model (Vieira et al. 2008). Similarly, *Lactobacillus sakei* 2a protected gnotobiotic mice against subsequent oral infection with *L. monocytogenes* (Bambirra et al. 2007).

The mechanisms by which potentially probiotic strains may impede colonization of pathogens have also been investigated. Competitive exclusion of pathogens by probiotic commensals has been suggested as a mechanism for displacing pathogens from the host cell surface (Collado et al. 2005). In addition, *Lactobacillus*-mediated stimulation of mucin production (*muc3*) by gut HT29 cells has been shown to protect against enteropathogenic *E. coli* (EPEC) (Mack et al. 2003). This work is particularly interesting in light of studies demonstrating that *L. monocytogenes* has the potential to interact with human Muc2 (Linden et al. 2008) however, the true role of the microbiota in inducing *muc2* expression has not yet been established. Furthermore, it has been demonstrated that *Lb. salivarius* UCC118 mediates protection against *L. monocytogenes* through bacteriocin production. Mice receiving orally administered wild-type UCC118 bacteria demonstrated significant protection against subsequent oral infection with *L. monocytogenes;* protection that was abolished in mice receiving a UCC118 mutant which lacks the ability to produce the bacteriocin. Furthermore, a bacteriocin-resistant variant of *L. monocytogenes* was immune to protection by wild-type UCC118. This work demonstrates that bacteriocin production represents a mechanism for probiotic/commensal-mediated protection against *L. monocytogenes* infection *in situ* within the GI tract (Corr et al. 2007b).

A study by Mark Lecuit and colleagues defined the intestinal host responses to *L. monocytogenes* infection in a gnotobiotic humanized (hEcad) mouse model (Lecuit et al. 2007). Transcriptional responses in ileal cells were compared across mice monocolonized by *L. monocytogenes*, *L. innocua* or the commensal *Bacteroides thetaiotaomicron*. In this setting *L. monocytogenes* initiated a profound immunoinflammatory response that was not induced by either *L. innocua* or *B. thetaiotaomicron.* This response included the stimulation of chemokines (Cxcl9 and Cxcl10) necessary for recruitment of T lymphocytes and identified TNFα as a central regulator of the response. Furthermore lack of host cell invasion (in Δ*inlA* and Δ*inlAB* mutants) or lack of cytoplasmic replication (in a Δ*hly* mutant) significantly reduced the ileal immunoinflammatory response even though luminal numbers of bacteria were comparable. This study revealed details of the cellular response evoked by *L. monocytogenes* infection of the small intestine and verified that the host response is specific to the pathogen and distinct from that induced by commensals or non-pathogens.

CONCLUSION

L. monocytogenes has long been utilised as a model intracellular pathogen with which to examine microbe-host interactions and the stimulation of specific anti-microbial immune responses (Cossart, 2007). However, work outlined in the current review demonstrates that the pathogen may also provide a tractable bacterial model with which to examine factors required for microbial survival in the gastrointestinal environment. It is clear, for example that bacterial factors such as BSH may be shared between pathogens (including *L. monocytogenes*) and the host microbiota (Jones et al. 2008). In addition, bacterial systems involved in adaptation to local microenvironments in the gut have been shown to play a role in transient colonisation by the pathogen (Sleator et al. 2001; Sleator ct al. 2005) and many of these are regulated by the central regulator of stress adaptation, σ^B (Chaturongakul et al. 2008; Hain et al. 2008). Many of the systems involved in stress adaptation also contribute to survival in the external environment and in the food matrix and therefore contribute to the entire infectious cycle (Gahan and Hill 2005).

Acknowledgments

RDS is a senior lecturer at the Department of Biological Sciences at Cork Institute of Technology and coordinator of the EU FP7 IAPP project ClouDx-i. This chapter is based on a previously published review article; Sleator RD, Watson D, Hill C, Gahan CG. The interaction between *Listeria monocytogenes* and the host gastrointestinal tract. Microbiology 2009; 155:2463-75.

References

Akerley, B. J., Cotter, P. A. & Miller, J. F. (1995). Ectopic expression of the flagellar regulon alters development of the Bordetella-host interaction. *Cell* 80, 611-620.

Altenhoefer, A., Oswald, S., Sonnenborn, U., Enders, C., Schulze, J., Hacker, J. & Oelschlaeger, T. A. (2004). The probiotic Escherichia coli strain Nissle 1917 interferes with invasion of human intestinal epithelial cells by different enteroinvasive bacterial pathogens. *FEMS Immunol Med Microbiol* 40, 223-229.

Andersen, J. B., Roldgaard, B. B., Lindner, A. B., Christensen, B. B. & Licht, T. R. (2006). Construction of a multiple fluorescence labelling system for use in co-invasion studies of Listeria monocytogenes. *BMC Microbiol* 6, 86.

Bahey-El-Din, M., Casey, P. G., Griffin, B. T. & Gahan, C. G. M. (2008). Lactococcus lactis-expressing listeriolysin O (LLO) provides protection and specific CD8(+) T cells against *Listeria monocytogenes* in the murine infection model. *Vaccine* 26, 5304-5314.

Bambirra, F. H., Lima, K. G., Franco, B. D., Cara, D. C., Nardi, R. M., Barbosa, F. H. & Nicoli, J. R. (2007). Protective effect of Lactobacillus sakei 2a against experimental challenge with *Listeria monocytogenes* in gnotobiotic mice. *Lett Appl Microbiol* 45, 663-667.

Becker, L. A., Cetin, M. S., Hutkins, R. W. & Benson, A. K. (1998). Identification of the gene encoding the alternative sigma factor sigmaB from *Listeria monocytogenes* and its role in osmotolerance. *J Bacteriol* 180, 4547-4554.

Becker, L. A., Evans, S. N., Hutkins, R. W. & Benson, A. K. (2000). Role of sigma(B) in adaptation of *Listeria monocytogenes* to growth at low temperature. *J Bacteriol* 182, 7083-7087.

Begley, M., Gahan, C. G. M. & Hill, C. (2002). Bile stress response in *Listeria monocytogenes* LO28: Adaptation, cross-protection, and identification of genetic loci involved in bile resistance. *Applied and Environmental Microbiology* 68, 6005-6012.

Begley, M., Sleator, R. D., Gahan, C. G. M. & Hill, C. (2005). Contribution of three bile-associated loci, bsh, pva, and btlB, to gastrointestinal persistence and bile tolerance of Listeria monocytogenes. *Infection and Immunity* 73, 894-904.

Begley, M., Bron, P. A., Heuston, S., Casey, P. G., Englert, N., Wiesner, J., Jomaa, H., Gahan, C. G. M. & Hill, C. (2008). Analysis of the Isoprenoid Biosynthesis Pathways in *Listeria monocytogenes* Reveals a Role for the Alternative 2-C-Methyl-D-Erythritol 4-Phosphate Pathway in Murine Infection. *Infection and Immunity* 76, 5392-5401.

Bonazzi, M., Lecuit, M. & Cossart, P. (2009). *Listeria monocytogenes* internalin and E-cadherin: from structure to pathogenesis. *Cell Microbiol.*

Brockstedt, D. G., Giedlin, M. A., Leong, M. L. & other authors (2004). Listeria-based cancer vaccines that segregate immunogenicity from toxicity. *Proc Natl Acad Sci U S A* 101, 13832-13837.

Brockstedt, D. G., Bahjat, K. S., Giedlin, M. A. & other authors (2005). Killed but metabolically active microbes: a new vaccine paradigm for eliciting effector T-cell responses and protective immunity. *Nat Med* 11, 853-860.

Bron, P. A., Monk, I. R., Corr, S. C., Hill, C. & Gahan, C. G. M. (2006). Novel luciferase reporter system for in vitro and organ-specific monitoring of differential gene expression in Listetia monocytogenes. *Applied and Environmental Microbiology* 72, 2876-2884.

Bublitz, M., Polle, L., Holland, C., Heinz, D. W., Nimtz, M. & Schubert, W. D. (2009). Structural basis for autoinhibition and activation of Auto, a virulence-associated peptidoglycan hydrolase of Listeria monocytogenes. *Mol Microbiol* 71, 1509-1522.

Buchrieser, C. (2007). Biodiversity of the species *Listeria monocytogenes* and the genus Listeria. *Microbes Infect* 9, 1147-1155.

Cabanes, D., Sousa, S., Cebria, A., Lecuit, M., Garcia-del Portillo, F. & Cossart, P. (2005). Gp96 is a receptor for a novel *Listeria monocytogenes* virulence factor, Vip, a surface protein. *EMBO J* 24, 2827-2838.

Cetin, M. S., Zhang, C., Hutkins, R. W. & Benson, A. K. (2004). Regulation of transcription of compatible solute transporters by the general stress sigma factor, sigmaB, in Listeria monocytogenes. *J Bacteriol* 186, 794-802.

Chaturongakul, S., Raengpradub, S., Wiedmann, M. & Boor, K. J. (2008). Modulation of stress and virulence in Listeria monocytogenes. *Trends Microbiol* 16, 388-396.

Collado, M. C., Gueimonde, M., Hernandez, M., Sanz, Y. & Salminen, S. (2005). Adhesion of selected Bifidobacterium strains to human intestinal mucus and the role of adhesion in enteropathogen exclusion. *J Food Prot* 68, 2672-2678.

Corr, S. C., Gahan, C. G. M. & Hill, C. (2007a). Impact of selected Lactobacillus and Bifidobacterium species on *Listeria monocytogenes* infection and the mucosal immune response. *Fems Immunology and Medical Microbiology* 50, 380-388.

Corr, S. C., Li, Y., Riedel, C. U., O'Toole, P. W., Hill, C. & Gahan, C. G. M. (2007b). Bacteriocin production as a mechanism for the antfinfective activity of Lactobacillus salivarius UCC118. *Proceedings of the National Academy of Sciences of the United States of America* 104, 7617-7621.

Cossart, P. (2007). Listeriology (1926-2007): the rise of a model pathogen. *Microbes Infect* 9, 1143-1146.

Cotter, P. D., Gahan, C. G. & Hill, C. (2001). A glutamate decarboxylase system protects *Listeria monocytogenes* in gastric fluid. *Mol Microbiol* 40, 465-475.

Cotter, P. D., Draper, L. A., Lawton, E. M., Daly, K. M., Groeger, D. S., Casey, P. G., Ross, R. P. & Hill, C. (2008). Listeriolysin S, a novel peptide haemolysin associated with a subset of lineage I Listeria monocytogenes. *PLoS Pathog* 4, e1000144.

Czuprynski, C. J. & Balish, E. (1981). Pathogenesis of *Listeria monocytogenes* for gnotobiotic rats. *Infect Immun* 32, 323-331.

Czuprynski, C. J., Faith, N. G. & Steinberg, H. (2002). Ability of the *Listeria monocytogenes* strain Scott A to cause systemic infection in mice infected by the intragastric route. *Appl Environ Microbiol* 68, 2893-2900.

Czuprynski, C. J. (2005). Listeria monocytogenes: silage, sandwiches and science. *Anim Health Res Rev* 6, 211-217.

Disson, O., Grayo, S., Huillet, E. & other authors (2008). Conjugated action of two species-specific invasion proteins for fetoplacental listeriosis. *Nature* 455, 1114-1118.

Dramsi, S., Bourdichon, F., Cabanes, D., Lecuit, M., Fsihi, H. & Cossart, P. (2004). FbpA, a novel multifunctional *Listeria monocytogenes* virulence factor. *Mol Microbiol* 53, 639-649.

Dussurget, O., Cabanes, D., Dehoux, P., Lecuit, M., Buchrieser, C., Glaser, P. & Cossart, P. (2002). *Listeria monocytogenes* bile salt hydrolase is a virulence factor involved in the intestinal and hepatic phases of listeriosis (vol 45, pg 1095, 2002). *Molecular Microbiology* 46, 903-903.

Ferreira, A., O'Byrne, C. P. & Boor, K. J. (2001). Role of sigma(B) in heat, ethanol, acid, and oxidative stress resistance and during carbon starvation in Listeria monocytogenes. *Appl Environ Microbiol* 67, 4454-4457.

Fraser, K. R., Sue, D., Wiedmann, M., Boor, K. & O'Byrne, C. P. (2003). Role of sigmaB in regulating the compatible solute uptake systems of Listeria monocytogenes: osmotic induction of opuC is sigmaB dependent. *Appl Environ Microbiol* 69, 2015-2022.

Fuhs, G. W. & Seeliger, H. P. (1961). [On flagellation of Listeria monocytogenes. Electronoptic and serological studies.]. *Arch Mikrobiol* 40, 153-162.

Gahan, C. G. & Hill, C. (2005). Gastrointestinal phase of *Listeria monocytogenes* infection. *J Appl Microbiol* 98, 1345-1353.

Garner, M. R., Njaa, B. L., Wiedmann, M. & Boor, K. J. (2006). Sigma B contributes to *Listeria monocytogenes* gastrointestinal infection but not to systemic spread in the guinea pig infection model. *Infect Immun* 74, 876-886.

Giron, J. A., Torres, A. G., Freer, E. & Kaper, J. B. (2002). The flagella of enteropathogenic Escherichia coli mediate adherence to epithelial cells. *Mol Microbiol* 44, 361-379.

Glaser, P., Frangeul, L., Buchrieser, C. & other authors (2001). Comparative genomics of Listeria species. *Science* 294, 849-852.

Gray, M. J., Freitag, N. E. & Boor, K. J. (2006). How the bacterial pathogen *Listeria monocytogenes* mediates the switch from environmental Dr. Jekyll to pathogenic Mr. Hyde. *Infection and Immunity* 74, 2505-2512.

Grundling, A., Burrack, L. S., Bouwer, H. G. & Higgins, D. E. (2004). *Listeria monocytogenes* regulates flagellar motility gene expression through MogR, a transcriptional repressor required for virulence. *Proc Natl Acad Sci U S A* 101, 12318-12323.

Guimaraes, V. D., Gabriel, J. E., Lefevre, F., Cabanes, D., Gruss, A., Cossart, P., Azevedo, V. & Langella, P. (2005). Internalin-expressing Lactococcus lactis is able to invade small intestine of guinea pigs and deliver DNA into mammalian epithelial cells. *Microbes Infect* 7, 836-844.

Hain, T., Steinweg, C., Kuenne, C. T. & other authors (2006). Whole-genome sequence of Listeria welshimeri reveals common steps in genome reduction with Listeria innocua as compared to Listeria monocytogenes. *Journal of Bacteriology* 188, 7405-7415.

Hain, T., Hossain, H., Chatterjee, S. S. & other authors (2008). Temporal transcriptomic analysis of the *Listeria monocytogenes* EGD-e sigmaB regulon. *BMC Microbiol* 8, 20.

Hamon, M., Bierne, H. & Cossart, P. (2006). Listeria monocytogenes: a multifaceted model. *Nat Rev Microbiol* 4, 423-434.

Hardy, J., Francis, K. P., DeBoer, M., Chu, P., Gibbs, K. & Contag, C. H. (2004). Extracellular replication of *Listeria monocytogenes* in the murine gall bladder. *Science* 303, 851-853.

Hu, Y., Oliver, H. F., Raengpradub, S., Palmer, M. E., Orsi, R. H., Wiedmann, M. & Boor, K. J. (2007a). Transcriptomic and phenotypic analyses suggest a network between the transcriptional regulators HrcA and sigmaB in Listeria monocytogenes. *Appl Environ Microbiol* 73, 7981-7991.

Hu, Y., Raengpradub, S., Schwab, U., Loss, C., Orsi, R. H., Wiedmann, M. & Boor, K. J. (2007b). Phenotypic and transcriptomic analyses demonstrate interactions between the transcriptional regulators CtsR and Sigma B in Listeria monocytogenes. *Appl Environ Microbiol* 73, 7967-7980.

Inagaki, H., Suzuki, T., Nomoto, K. & Yoshikai, Y. (1996). Increased susceptibility to primary infection with *Listeria monocytogenes* in germfree mice may be due to lack of accumulation of L-selectin+ CD44+ T cells in sites of inflammation. *Infect Immun* 64, 3280-3287.

Ireton, K. (2007). Entry of the bacterial pathogen *Listeria monocytogenes* into mammalian cells. *Cell Microbiol* 9, 1365-1375.

Jacquet C, Doumith M, Gordon JI, Martin PM, Cossart P, Lecuit M. (2004). A molecular marker for evaluating the pathogenic potential of foodborne Listeria monocytogenes. J Infect Dis. 189, 2094-2100.

Jeffers, G. T., Bruce, J. L., McDonough, P. L., Scarlett, J., Boor, K. J. & Wiedmann, M. (2001). Comparative genetic characterization of *Listeria monocytogenes* isolates from human and animal listeriosis cases. *Microbiology* 147, 1095-1104.

Jones, B. V., Begley, M., Hill, C., Gahan, C. G. & Marchesi, J. R. (2008). Functional and comparative metagenomic analysis of bile salt hydrolase activity in the human gut microbiome. *Proc Natl Acad Sci U S A* 105, 13580-13585.

Kazmierczak, M. J., Mithoe, S. C., Boor, K. J. & Wiedmann, M. (2003). *Listeria monocytogenes* sigma B regulates stress response and virulence functions. *J Bacteriol* 185, 5722-5734.

Khelef, N., Lecuit, M., Bierne, H. & Cossart, P. (2006). Species specificity of the *Listeria monocytogenes* InlB protein. *Cell Microbiol* 8, 457-470.

Kim, K. P., Jagadeesan, B., Burkholder, K. M., Jaradat, Z. W., Wampler, J. L., Lathrop, A. A., Morgan, M. T. & Bhunia, A. K. (2006). Adhesion characteristics of Listeria adhesion protein (LAP)-expressing Escherichia coli to Caco-2 cells and of recombinant LAP to eukaryotic receptor Hsp60 as examined in a surface plasmon resonance sensor. *FEMS Microbiol Lett* 256, 324-332.

Kim, S. H., Bakko, M. K., Knowles, D. & Borucki, M. K. (2004). Oral inoculation of A/J mice for detection of invasiveness differences between *Listeria monocytogenes* epidemic and environmental strains. *Infect Immun* 72, 4318-4321.

Lecuit, M., Vandormael-Pournin, S., Lefort, J., Huerre, M., Gounon, P., Dupuy, C., Babinet, C. & Cossart, P. (2001). A transgenic model for listeriosis: role of internalin in crossing the intestinal barrier. *Science* 292, 1722-1725.

Lecuit, M. (2007). Human listeriosis and animal models. *Microbes Infect* 9, 1216-1225.

Lecuit, M., Sonnenburg, J. L., Cossart, P. & Gordon, J. I. (2007). Functional genomic studies of the intestinal response to a foodborne enteropathogen in a humanized gnotobiotic mouse model. *J Biol Chem* 282, 15065-15072.

Linden, S. K., Bierne, H., Sabet, C., Png, C. W., Florin, T. H., McGuckin, M. A. & Cossart, P. (2008). *Listeria monocytogenes* internalins bind to the human intestinal mucin MUC2. *Arch Microbiol* 190, 101-104.

Maa, Y. F. & Prestrelski, S. J. (2000). Biopharmaceutical powders: particle formation and formulation considerations. *Curr Pharm Biotechnol* 1, 283-302.

Mack, D. R., Ahrne, S., Hyde, L., Wei, S. & Hollingsworth, M. A. (2003). Extracellular MUC3 mucin secretion follows adherence of Lactobacillus strains to intestinal epithelial cells in vitro. *Gut* 52, 827-833.

Mengaud, J., Lecuit, M., Lebrun, M., Nato, F., Mazie, J. C. & Cossart, P. (1996). Antibodies to the leucine-rich repeat region of internalin block entry of *Listeria monocytogenes* into cells expressing E-cadherin. *Infect Immun* 64, 5430-5433.

Milillo, S. R., Badamo, J. M. & Wiedmann, M. (2009). Contributions to selected phenotypic characteristics of large species- and lineage-specific genomic regions in Listeria monocytogenes. *Food Microbiol* 26, 212-223.

Monk, I. R., Casey, P. G., Cronin, M., Gahan, C. G. & Hill, C. (2008a). Development of multiple strain competitive index assays for *Listeria monocytogenes* using pIMC; a new site-specific integrative vector. *Bmc Microbiology* 8.

Monk, I. R., Gahan, C. G. M. & Hill, C. (2008b). Tools for functional postgenomic analysis of Listeria monocytogenes. *Applied and Environmental Microbiology* 74, 3921-3934.

Moorhead, S. M. & Dykes, G. A. (2003). The role of the sigB gene in the general stress response of *Listeria monocytogenes* varies between a strain of serotype 1/2a and a strain of serotype 4c. *Curr Microbiol* 46, 461-466.

Nadon, C. A., Bowen, B. M., Wiedmann, M. & Boor, K. J. (2002). Sigma B contributes to PrfA-mediated virulence in Listeria monocytogenes. *Infect Immun* 70, 3948-3952.

Nelson, K. E., Fouts, D. E., Mongodin, E. F. & other authors (2004). Whole genome comparisons of serotype 4b and 1/2a strains of the food-borne pathogen *Listeria monocytogenes* reveal new insights into the core genome components of this species. *Nucleic Acids Research* 32, 2386-2395.

Nightingale, K. K., Windham, K. & Wiedmann, M. (2005). Evolution and molecular phylogeny of *Listeria monocytogenes* isolated from human and animal listeriosis cases and foods. *J Bacteriol* 187, 5537-5551.

O'Neil, H. S. & Marquis, H. (2006). *Listeria monocytogenes* flagella are used for motility, not as adhesins, to increase host cell invasion. *Infect Immun* 74, 6675-6681.

Ollinger, J., Bowen, B., Wiedmann, M., Boor, K. J. & Bergholz, T. M. (2009). *Listeria monocytogenes* {sigma}B modulates PrfA-mediated virulence factor expression. *Infect Immun*.

Peel, M., Donachie, W. & Shaw, A. (1988). Temperature-dependent expression of flagella of *Listeria monocytogenes* studied by electron microscopy, SDS-PAGE and western blotting. *J Gen Microbiol* 134, 2171-2178.

Pentecost, M., Otto, G., Theriot, J. A. & Amieva, M. R. (2006). *Listeria monocytogenes* invades the epithelial junctions at sites of cell extrusion. *PLoS Pathog* 2, e3.

Raengpradub, S., Wiedmann, M. & Boor, K. J. (2008). Comparative analysis of the sigma B-dependent stress responses in *Listeria monocytogenes* and Listeria innocua strains exposed to selected stress conditions. *Appl Environ Microbiol* 74, 158-171.

Rafelski, S. M. & Theriot, J. A. (2006). Mechanism of polarization of *Listeria monocytogenes* surface protein ActA. *Mol Microbiol* 59, 1262-1279.

Ragon, M., Wirth, T., Hollandt, F., Lavenir, R., Lecuit, M., Le Monnier, A. & Brisse, S. (2008). A new perspective on Listeria monocytogenes evolution. *PLoS Pathog* 4, e1000146.

Rastall, R. A. (2004). Bacteria in the gut: friends and foes and how to alter the balance. *J Nutr* 134, 2022S-2026S.

Rauch, M., Luo, Q., Muller-Altrock, S. & Goebel, W. (2005). SigB-dependent in vitro transcription of prfA and some newly identified genes of *Listeria monocytogenes* whose expression is affected by PrfA in vivo. *J Bacteriol* 187, 800-804.

Riedel, C. U., Monk, I. R., Casey, P. G., Morrissey, D., O'Sullivan, G. C., Tangney, M., Hill, C. & Gahan, C. G. M. (2007). Improved luciferase tagging system for *Listeria monocytogenes* allows real-time monitoring in vivo and in vitro. *Applied and Environmental Microbiology* 73, 3091-3094.

Roland, K. L., Tinge, S. A., Killeen, K. P. & Kochi, S. K. (2005). Recent advances in the development of live, attenuated bacterial vectors. *Curr Opin Mol Ther* 7, 62-72.

Ryan, S., Hill, C. & Gahan, C. G. (2008). Acid stress responses in Listeria monocytogenes. *Adv Appl Microbiol* 65, 67-91.

Ryan, S., Begley, M., Gahan, C. G. M. & Hill, C. (2009). Molecular characterization of the arginine deiminase system in Listeria monocytogenes: regulation and role in acid tolerance. *Environmental Microbiology* 11, 432-445.

Sabet, C., Toledo-Arana, A., Personnic, N. & other authors (2008). The *Listeria monocytogenes* virulence factor InlJ is specifically expressed in vivo and behaves as an adhesin. *Infect Immun* 76, 1368-1378.

Schwab, U., Bowen, B., Nadon, C., Wiedmann, M. & Boor, K. J. (2005). The *Listeria monocytogenes* prfAP2 promoter is regulated by sigma B in a growth phase dependent manner. *FEMS Microbiol Lett* 245, 329-336.

Seegers, J. F. (2002). Lactobacilli as live vaccine delivery vectors: progress and prospects. *Trends Biotechnol* 20, 508-515.

Seveau, S., Pizarro-Cerda, J. & Cossart, P. (2007). Molecular mechanisms exploited by *Listeria monocytogenes* during host cell invasion. *Microbes Infect* 9, 1167-1175.

Shahabi, V., Reyes-Reyes, M., Wallecha, A., Rivera, S., Paterson, Y. & Maciag, P. (2008). Development of a *Listeria monocytogenes* based vaccine against prostate cancer. *Cancer Immunol Immunother* 57, 1301-1313.

Shahidi, F. & Han, X. Q. (1993). Encapsulation of food ingredients. *Crit Rev Food Sci Nutr* 33, 501-547.

Sheehan, V. M., Sleator, R. D., Fitzgerald, G. F. & Hill, C. (2006). Heterologous expression of BetL, a betaine uptake system, enhances the stress tolerance of Lactobacillus salivarius UCC118. *Appl Environ Microbiol* 72, 2170-2177.

Sheehan, V. M., Sleator, R. D., Hill, C. & Fitzgerald, G. F. (2007). Improving gastric transit, gastrointestinal persistence and therapeutic efficacy of the probiotic strain Bifidobacterium breve UCC2003. *Microbiology* 153, 3563-3571.

Shen, A. & Higgins, D. E. (2006). The MogR transcriptional repressor regulates nonhierarchal expression of flagellar motility genes and virulence in Listeria monocytogenes. *PLoS Pathog* 2, e30.

Sleator, R. D., Wouters, J., Gahan, C. G. M., Abee, T. & Hill, C. (2001). Analysis of the role of OpuC, an osmolyte transport system, in salt tolerance and virulence potential of Listeria monocytogenes. *Applied and Environmental Microbiology* 67, 2692-2698.

Sleator, R. D. & Hill, C. (2002). Bacterial osmoadaptation: the role of osmolytes in bacterial stress and virulence. *FEMS Microbiol Rev* 26, 49-71.

Sleator, R. D., Francis, G. A., O'Beirne, D., Gahan, C. G. & Hill, C. (2003a). Betaine and carnitine uptake systems in *Listeria monocytogenes* affect growth and survival in foods and during infection. *J Appl Microbiol* 95, 839-846.

Sleator, R. D., Gahan, C. G. & Hill, C. (2003b). A postgenomic appraisal of osmotolerance in Listeria monocytogenes. *Appl Environ Microbiol* 69, 1-9.

Sleator, R. D., Wemekamp-Kamphuis, H. H., Gahan, C. G. M., Abee, T. & Hill, C. (2005). A PrfA-regulated bile exclusion system (BilE) is a novel virulence factor in Listeria monocytogenes. *Molecular Microbiology* 55, 1183-1195.

Sleator, R. D. & Hill, C. (2006). Patho-biotechnology: using bad bugs to do good things. *Curr Opin Biotechnol* 17, 211-216.

Sleator, R. D., Clifford, T. & Hill, C. (2007). Gut osmolarity: a key environmental cue initiating the gastrointestinal phase of *Listeria monocytogenes* infection? *Med Hypotheses* 69, 1090-1092.

Sleator, R. D. & Hill, C. (2007). Patho-biotechnology; using bad bugs to make good bugs better. *Sci Prog* 90, 1-14.

Sleator, R. D. & Hill, C. (2008a). 'Bioengineered Bugs' - a patho-biotechnology approach to probiotic research and applications. *Med Hypotheses* 70, 167-169.

Sleator, R. D. & Hill, C. (2008b). Battle of the bugs. *Science* 321, 1294-1295.

Sue, D., Fink, D., Wiedmann, M. & Boor, K. J. (2004). sigmaB-dependent gene induction and expression in *Listeria monocytogenes* during osmotic and acid stress conditions simulating the intestinal environment. *Microbiology* 150, 3843-3855.

Swaminathan, B. & Gerner-Smidt, P. (2007). The epidemiology of human listeriosis. *Microbes and Infection* 9, 1236-1243.

Vazquez-Boland, J. A., Kuhn, M., Berche, P., Chakraborty, T., Dominguez-Bernal, G., Goebel, W., Gonzalez-Zorn, B., Wehland, J. & Kreft, J. (2001). Listeria pathogenesis and molecular virulence determinants. *Clin Microbiol Rev* 14, 584-640.

Vieira, L. Q., dos Santos, L. M., Neumann, E., da Silva, A. P., Moura, L. N. & Nicoli, J. R. (2008). Probiotics protect mice against experimental infections. *J Clin Gastroenterol* 42 Suppl 3 Pt 2, S168-169.

Volker, U., Maul, B. & Hecker, M. (1999). Expression of the sigmaB-dependent general stress regulon confers multiple stress resistance in Bacillus subtilis. *J Bacteriol* 181, 3942-3948.

Wallecha, A., Maciag, P. C., Rivera, S., Paterson, Y. & Shahabi, V. (2009). Construction and characterization of an attenuated *Listeria monocytogenes* strain for clinical use in cancer immunotherapy. *Clin Vaccine Immunol* 16, 96-103.

Wampler, J. L., Kim, K. P., Jaradat, Z. & Bhunia, A. K. (2004). Heat shock protein 60 acts as a receptor for the Listeria adhesion protein in Caco-2 cells. *Infect Immun* 72, 931-936.

Watson, D., Sleator, R. D., Hill, C. & Gahan, C. G. M. (2008). Enhancing bile tolerance improves survival and persistence of Bifidobacterium and Lactococcus in the murine gastrointestinal tract. *Bmc Microbiology* 8.

Wemekamp-Kamphuis, H. H., Wouters, J. A., Sleator, R. D., Gahan, C. G. M., Hill, C. & Abee, T. (2002). Multiple deletions of the osmolyte transporters BetL, Gbu, and OpuC of *Listeria monocytogenes* affect virulence and growth at high osmolarity. *Applied and Environmental Microbiology* 68, 4710-4716.

Wemekamp-Kamphuis, H. H., Wouters, J. A., de Leeuw, P. P., Hain, T., Chakraborty, T. & Abee, T. (2004). Identification of sigma factor sigma B-controlled genes and their impact on acid stress, high hydrostatic pressure, and freeze survival in *Listeria monocytogenes* EGD-e. *Appl Environ Microbiol* 70, 3457-3466.

Wiedmann, M., Arvik, T. J., Hurley, R. J. & Boor, K. J. (1998). General stress transcription factor sigmaB and its role in acid tolerance and virulence of Listeria monocytogenes. *J Bacteriol* 180, 3650-3656.

Wiedmann, M. (2002). Molecular subtyping methods for Listeria monocytogenes. *J AOAC Int* 85, 524-531.

Wolfgang, M. C., Jyot, J., Goodman, A. L., Ramphal, R. & Lory, S. (2004). Pseudomonas aeruginosa regulates flagellin expression as part of a global response to airway fluid from cystic fibrosis patients. *Proc Natl Acad Sci U S A* 101, 6664-6668.

Wollert, T., Pasche, B., Rochon, M., Deppenmeier, S., van den Heuvel, J., Gruber, A. D., Heinz, D. W., Lengeling, A. & Schubert, W. D. (2007). Extending the host range of *Listeria monocytogenes* by rational protein design. *Cell* 129, 891-902.

Xu, J. & Gordon, J. I. (2003). Inaugural Article: Honor thy symbionts. *Proc Natl Acad Sci U S A* 100, 10452-10459.

In: Listeria monocytogenes
Editor: Tiffany Vicario
ISBN: 978-1-63483-765-1

Chapter 8

SEVERAL METHODS FOR DETERMINATION OF PHAGOCYTIC AND KILLING ACTIVITIES OF MACROPHAGES AGAINST *LISTERIA MONOCYTOGENES* INFECTION

WHICH IS THE SIMPLEST AND REPRODUCIBLE METHOD?

Masakazu Kaneko, Yu-ju Kanayama, Yoshiko Emoto and Masashi Emoto*
[1]Laboratory of Immunology, Department of Laboratory Sciences,
Gunma University Graduate School of Health Sciences,
Maebashi, Gunma, Japan

ABSTRACT

Listeria monocytogenes is a facultative intracellular bacterium and hence, this bacterium is able to survive and propagate in macrophages (Mϕ). Yet, in contrast to this notion, a majority of *L. monocytogenes* is killed by Mϕ, indicating that Mϕ play a critical role in protection against *L. monocytogenes* infection. Because the number of *L. monocytogenes* engulfed and killed by Mϕ depends on whether or not Mϕ are activated, experimental listeriosis has been employed for analyzing Mϕ activities by many investigators. There are several methods for analyzing phagocytic and listericidal activities of Mϕ. Yet, in most cases expensive materials and complicated methods are required. In this review, in addition to methods recently employed for determination of phagocytic and listericidal activities of Mϕ, simple and excellent, though old-fashioned, method is also introduced.

---

* E-mail: masashi.emoto@gmail.com; Tel.: +81-27-220-8935; fax: +81-27-220-8935.

INTRODUCTION

Listeria monocytogenes is a Gram-positive bacterial pathogen, which causes a serious and life-threatening food-borne disease [1]. Listeriosis is one of the most common lethal food-borne infectious diseases in the United States [2] and in the United Kingdom [3].

Adaptive immunity mainly played by $CD8^+$ T cells with cytotoxic and interferon-γ - secreting activities is essential for eradication of *L. monocytogenes* [4, 5]. Yet, innate immunity orchestrated by professional phagocytes such as macrophages (Mϕ) and polymorphonuclear leukocytes also plays a critical role in protection against *L. monocytogenes* infection as a first line of defense [6-10].

In general, after bacterial pathogens are engulfed by Mϕ, phagosomes are formatted therein [11]. The phagosome encounters lysosome comprising various anti-microbial agents such as lysozyme located in the cytosol and matures into phagolysosome [11]. Nitric oxide synthesized by inducible nitric oxide synthase and reactive oxygen intermediates play a critical role in killing of bacteria within the phagosome [12].

Bacteria are categorized into at least 2 groups on the basis of their kinetics in Mϕ [13], one of which is extracellular bacteria and the other intracellular bacteria. Although extracellular bacteria are easily killed by Mϕ, intracellular bacteria such as *L. monocytogenes* show resistance to attack by Mϕ.

L. monocytogenes has a unique self-defense mechanism against digestion by Mϕ; i.e., after engulfed by Mϕ, *L. monocytogenes* escapes from the phagosome into the cytosol by means of listeriolysin O (LLO) which is a major virulence factor of this bacterium [14, 15]. Thus, *L. monocytogenes* is able to survive and propagate in Mϕ using LLO. Yet, a vast majority of *L. monocytogenes* are actually killed by Mϕ, although it depends on whether or not Mϕ are activated [16, 17].

Because experimental listeriosis has been employed for analyzing Mϕ activities by many investigators, analysis of phagocytic and listericidal activities of Mϕ is a useful tool for determination of their functional activities. Although several methods are employed for determination of phagocytic and listericidal activities of Mϕ, expensive materials, expensive equipment and complicated methods are required. In this review, in addition to a couple of methods recently employed for determination of phagocytic and listericidal activities of Mϕ, an excellent and a simple, though old-fashioned, method for determination of phagocytic and listericidal activities of Mϕ which are simultaneously measurable is introduced.

A COUPLE OF METHODS RECENTLY EMPLOYED FOR DETERMINATION OF PHAGOCYTIC AND LISTERICIDAL ACTIVITIES OF Mϕ

L. monocytogenes engulfed by Mϕ can be observed under light and fluorescence microscopes (Figure 1A) [18, 19]. Although *L. monocytogenes* engulfed by Mϕ can also be observed, though indirectly, by flow cytometry (Figure 1A) [20], additional experiment should be performed to distinguish bacteria engulfed by Mϕ from those merely attached to their plasma membrane; i.e., *L. monocytogenes* engulfed by Mϕ should be distinguished from those attached to their plasma membrane. This can be performed by ethidium bromide (Figure 1B) [18, 19]. Although the phagocytic activity of Mϕ can be determined by these methods, another experiment must be performed to determine listericidal activity of Mϕ using saponin and MTT in each case (Figure 1C) [19, 21]. Although it is physically possible to determine phagocytic and listericidal activities of Mϕ by these methods, the materials and the equipment are expensive and the methods are rather complicated. Moreover, it is impossible to measure both phagocytic and listericidal activities of Mϕ simultaneously by these methods.

Simple and Excellent, though Old-Fashioned, Method for Determination of Phagocytic and Listericidal Activities of Mϕ

Phagocytic and listericidal activities of Mϕ can also be determined by counting colony-forming units (CFU) in Mϕ. The phagocytic activity of Mϕ can be determined by counting

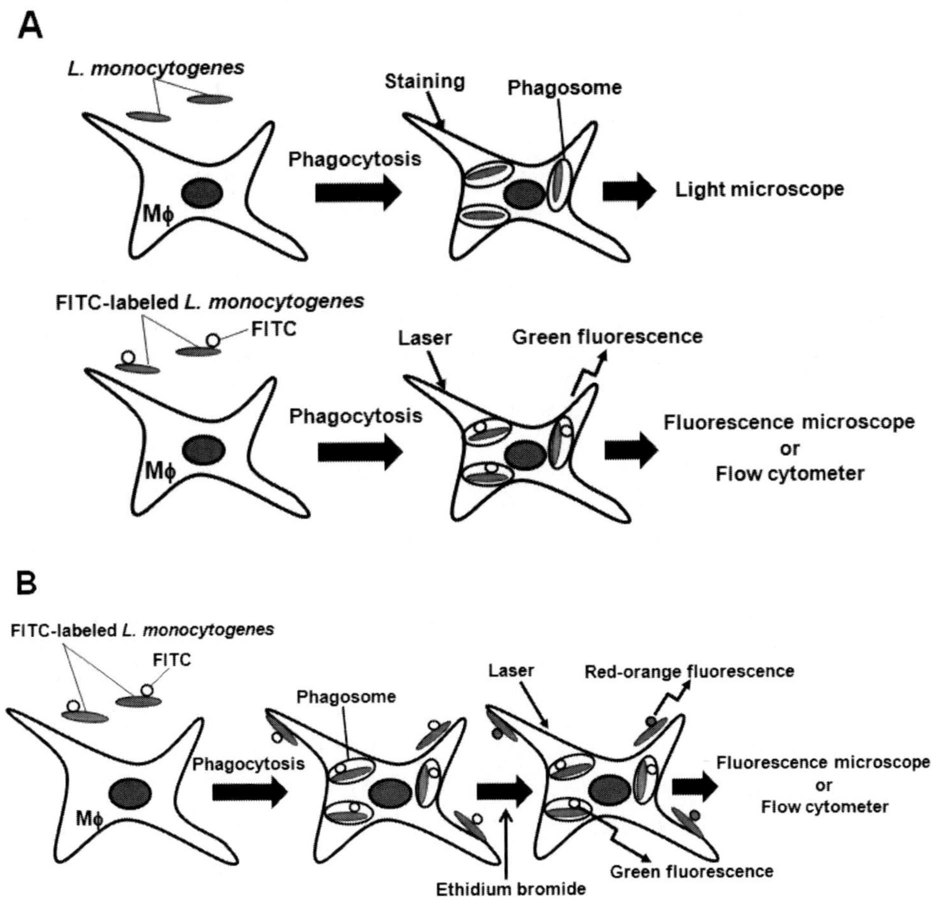

Figure 1. (Continued).

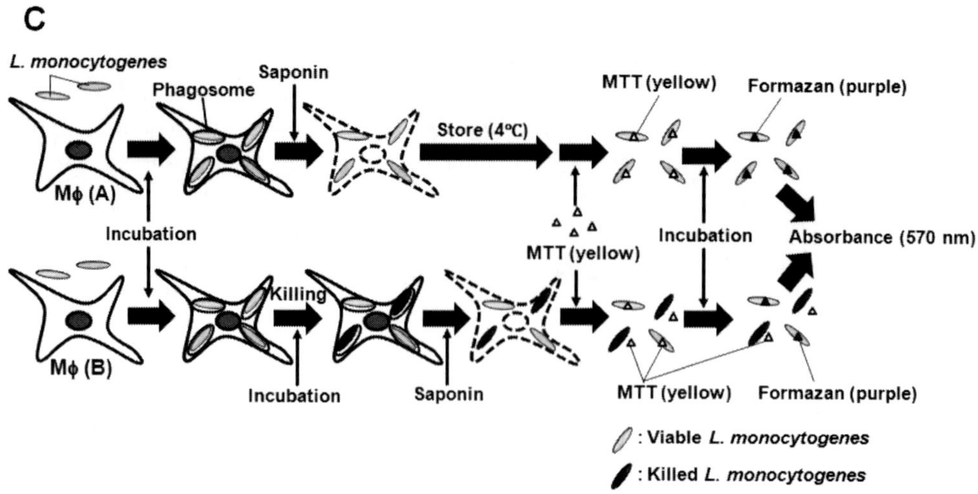

Figure 1. A couple of methods recently employed for determination of phagocytic and listericidal activities of Mϕ. (A) *L. monocytogenes* engulfed by Mϕ can be observed under light microscope (upper panel). FITC-labeled *L. monocytogenes* engulfed by Mϕ can be observed under fluorescence microscope (lower panel) and by flow cytometer (lower panel). (B) After incubation of Mϕ with *L. monocytogenes*, some bacteria which are not engulfed by, though attached to, Mϕ exist. To distinguish each bacterium, cells are stained with ethidium bromide. Extracellular *L. monocytogenes* (i.e., bacteria attached to the surface of Mϕ) fluorescence red-orange, whereas internalized *L. monocytogenes* (i.e., bacteria engulfed by Mϕ) fluorescence green. (C) Two sets of Mϕ (designated as Mϕ (A) and Mϕ (B)) are incubated with *L. monocytogenes* to be ingested by Mϕ. Mϕ (A) are treated with saponin and store at 4℃. *L. monocytogenes* released from Mϕ (A) are incubated with MTT and the absorbance is measured by absorption spectrometer at 570 nm (Absorbance A). Mϕ (B) infected with *L. monocytogenes* are further incubated at 37 °C after being ingested. During this period, *L. monocytogenes* engulfed by Mϕ are killed. To release *L. monocytogenes* from Mϕ, Mϕ (B) are treated with saponin. *L. monocytogenes* released from Mϕ (B) are incubated with MTT and the absorbance is measured (Absorbance B). After incubation with MTT, MTT (yellow color) is incorporated into *L. monocytogenes* and then become formazan (purple color) by dehydrogenase in mitocondria, which can be detected by absorption spectrometer at 570 nm. The listericidal activity is calculated as follows: 100 – {(Absorbance B / Absorbance A) × 100} (%).

CFU in Mϕ after short time incubation with *L. monocytogenes* (Figure 2A). After engulfed by Mϕ, *L. monocytogenes* is immediately killed by Mϕ and hence, listericidal activities of Mϕ can be determined by counting CFU in Mϕ after further incubation for a short time period (Figure 2A). The culture medium containing gentamicin (GM) must be used for washing out *L. monocytogenes* which are not engulfed by Mϕ. Although similar to other methods described above, preliminary experiments must be performed. This method is quite simple, and expensive materials and equipment are unnecessary. Moreover, phagocytic and listericidal activities can be determined simultaneously by this method. Accordingly, we assume that this method is the simplest and reproducible method for determination of phagocytic and listericidal activities of Mϕ. Therefore, we focus on this method (designed as CFU-counting method) in the following sections.

Determination of Phagocytic Activity of Mϕ by CFU-Counting Method

The CFU-counting method for determination of phagocytic activity of Mϕ is briefly described (Figure 2B). Mϕ are incubated with *L. monocytogenes* for 45 min to ingest bacteria.

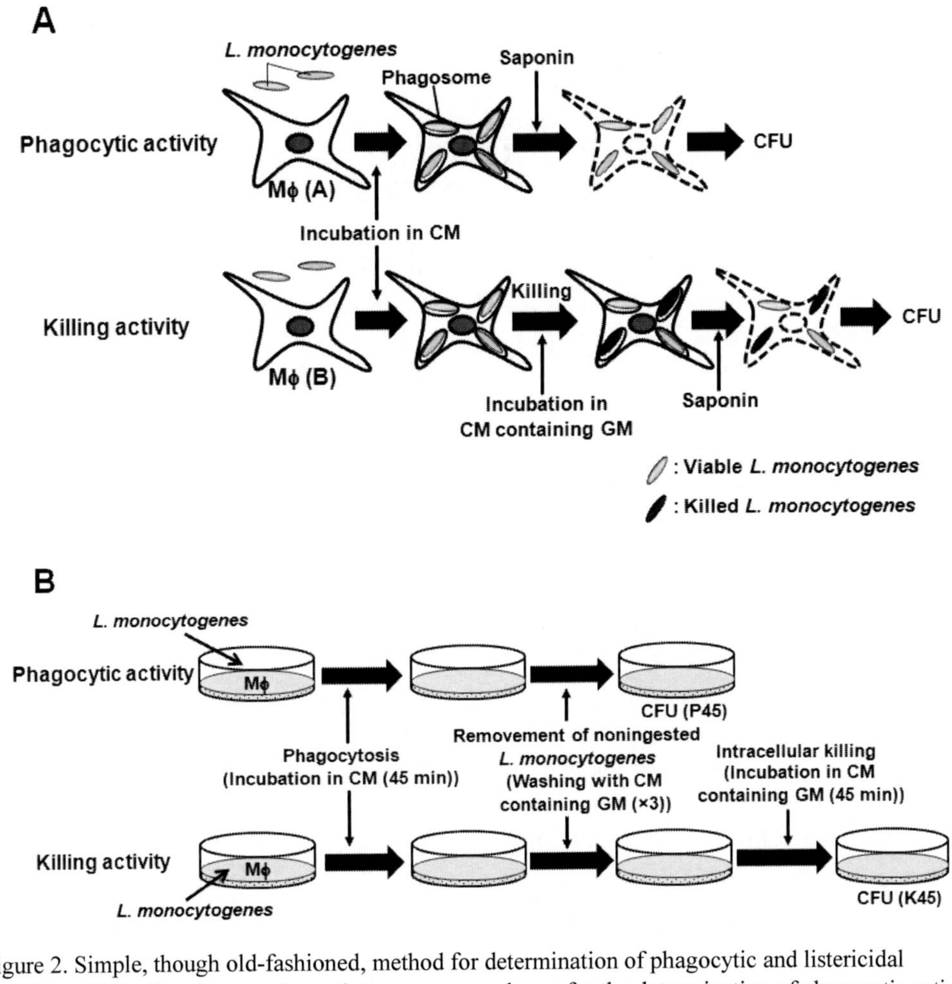

Figure 2. Simple, though old-fashioned, method for determination of phagocytic and listericidal activities of Mϕ. Two tissue culture plates are prepared; one for the determination of phagocytic activity (upper panel) and the other for the listericidal activity (lower panel). (A) Two sets of Mϕ are incubated with *L. monocytogenes* in CM to be ingested by Mϕ. For determination of phagocytic activity, Mϕ are treated with saponin and then CFU are determined. For determination of listericidal activity, Mϕ infected with *L. monocytogenes* are further incubated in CM containing GM. During this period, *L. monocytogenes* engulfed by Mϕ are killed. To release *L. monocytogenes* from Mϕ, Mϕ are treated with saponin and then CFU are determined. (B) Mϕ are incubated with *L. monocytogenes* (Mϕ : *L. monocytogenes* = 1 : 10) in CM for 45 min to ingest *L. monocytogenes*. Subsequently, cells are washed 3 times with CM containing GM (2.5-fold MIC) to remove noningested *L. monocytogenes* followed by CM, and then CFU are determined (phagocytic activity (P45)). Infected Mϕ in another plate are further incubated in CM containing GM (2.5-fold MIC) for 45 min to kill *L. monocytogenes* by Mϕ, and then CFU are determined (killing activity (K45)).

Subsequently, cells are washed 3 times with RPMI 1640 containing 10% fetal calf serum (designated as complete medium (CM)) and GM (2.5-fold minimal inhibitory concentrations (MIC)) to remove noningested bacteria followed by CM. Note that pre-warmed, but not cold, CM must be used throughout the experiment not to detach Mϕ from the bottom of tissue culture plates. A portion of cells is treated with saponin, plated on tryptic soy agar plates after sonication and the CFU are determined. Phagocytic activity is calculated as follows: {numbers of viable *L. monocytogenes* ingested by Mϕ (designated as P45) / numbers of *L. monocytogenes* incubated with Mϕ (designated as T_0)} ×100 (%). Thus, the percentage of *L. monocytogenes* engulfed by Mϕ can easily be quantitated by this method.

Determination of Listericidal Activity of Mϕ by CFU-Counting Method

The CFU-counting method for determination of listericidal activity of Mϕ is briefly described (Figure 2B). Mϕ are incubated with *L. monocytogenes* for 45 min and then washed 3 times with CM containing GM (2.5-fold MIC) to remove noningested bacteria. Subsequently, cells are further incubated in CM containing GM (2.5-fold MIC) for 45 min followed by CM, treated with saponin, plated on tryptic soy agar plates after sonication, and the CFU are determined. Listericidal activity is calculated as follows: 100 - {(numbers of remaining viable bacteria in Mϕ (designed as K45) / P45) × 100} (%). Thus, the percentage of viable *L. monocytogenes* in Mϕ can easily be quantitated by this method.

Why Do We Have to Employ GM?

One of the important points for the CFU-counting method is to eliminate extracellular *L. monocytogenes*. For killing uninfected *L. monocytogenes*, it is important to employ antibiotics which can kill extracellular, but not intracellular, bacteria. Antibiotics with low molecular weight penetrate into cytosol through plasma membrane of Mϕ, by which bacteria engulfed by Mϕ are killed. Therefore, the antibiotics with low-molecular weight cannot be used. In contrast, antibiotics with high molecular weight are unable to penetrate into cytosol and hence, bacteria engulfed by Mϕ are not killed. Although *L. monocytogenes* is susceptible to several antibiotics [22], most of them are low molecular weight [23]. GM is an aminoglycoside antibiotic which binds to bacterial ribosome 30s subunit and induces misreading of a wide range of RNAs [24]. GM is considered to be unable to penetrate into cytosol [25-29], because this antibiotic is a high molecular weight [23]. Accordingly, GM is recommended for the CFU-counting method.

Preliminary Experiments

The preliminary experiments which should be performed before the determination of phagocytic and listericidal activities of Mϕ are briefly described. It is important to determine optimal condition (i.e., concentration and length of effectiveness) of GM in advance.

Although it has been considered that GM cannot penetrate into cytosol [25-29], some studies report that GM, though high concentrations (i.e., > 10 μg/ml) and long-term incubation (i.e., > 2 h), can penetrate into cytosol through plasma membrane and kill *L. monocytogenes* in Mϕ [30, 31]. Moreover, we find that Mϕ are destroyed by high concentration (i.e., > 100 μg/ml) of GM (Kaneko, unpublished observation). Because MIC of GM is very low [22, 32] and because long-term incubation is unnecessary [22], *L. monocytogenes* in Mϕ are not killed by GM used for this experiment. However, because it should be verified whether extracellular, but not intracellular, *L. monocytogenes* is indeed killed by GM, the optimal concentration (i.e., MIC) and the length of effectiveness of GM used for determination of phagocytic and listericidal activities of Mϕ must be determined with care.

Culture medium (e.g., tryptic soy broth) containing different concentrations of GM should be prepared (Figure 3). After incubation of *L. monocytogenes* in culture medium containing different concentrations of GM, MIC is determined. In our laboratory, it appears that MIC of GM against *L. monocytogenes* is 2 μg/ml, and 2.5-fold MIC (i.e., 5 μg/ml) is suitable for killing extracellular, but not intracellular, *L. monocytogenes* in a short time period (Kaneko, unpublished observation). To determine the length of effectiveness of GM against *L. monocytogenes*, *L. monocytogenes* (1×10^5 CFU) is incubated in CM containing 5 μg/ml of GM (2.5-fold MIC) and the CFU are determined every 5 min (Figure 4). Viable *L. monocytogenes* become undetectable 20 min after incubation. To verify whether GM does not penetrate into cytosol, Mϕ incubated with *L. monocytogenes* for 45 min are washed 3 times with CM containing 5 μg/ml of GM (2.5-fold MIC) to remove noningested bacteria (Figure 5). Cells are further incubated in CM containing 5 μg/ml of GM (2.5-fold MIC) for 45 min, treated with 0.5% saponin, plated on agar plates after sonication, and the CFU are determined.

For determination of CFU in Mϕ, Mϕ should be destroyed to release internalized *L. monocytogenes*. Saponin facilitates destroy of plasma and phagosomal membrane of Mϕ by interacting with a cholesterol which is abundant in the plasma membrane [33]. To exclude the possibility that *L. monocytogenes* is killed by saponin, optimal condition (i.e., concentration and length of effectiveness) of saponin should also be determined in advance. In our laboratory, it appears that 0.5% saponin does not affect the viability of *L. monocytogenes* (Kaneko, unpublished observation).

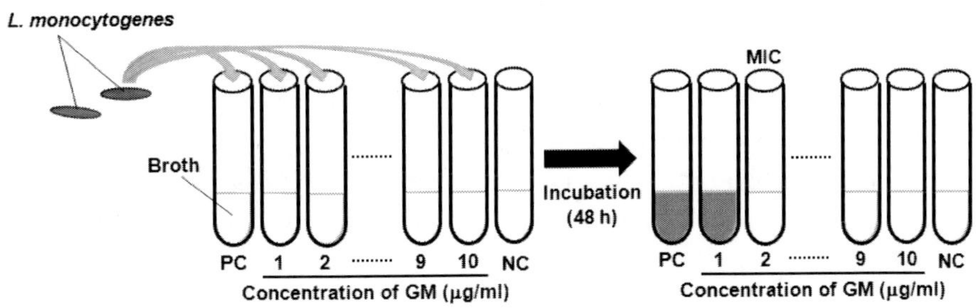

Figure 3. Experimental procedure for determination of MIC. *L. monocytogenes* is incubated in culture broth with or without various concentrations of GM at 37°C. Positive and negative controls (PC and NC) should also be prepared. Turbid and clear tube(s) represent bacterial growth and growth inhibition, respectively. The lowest concentration of GM inhibiting listerial growth is regarded as MIC.

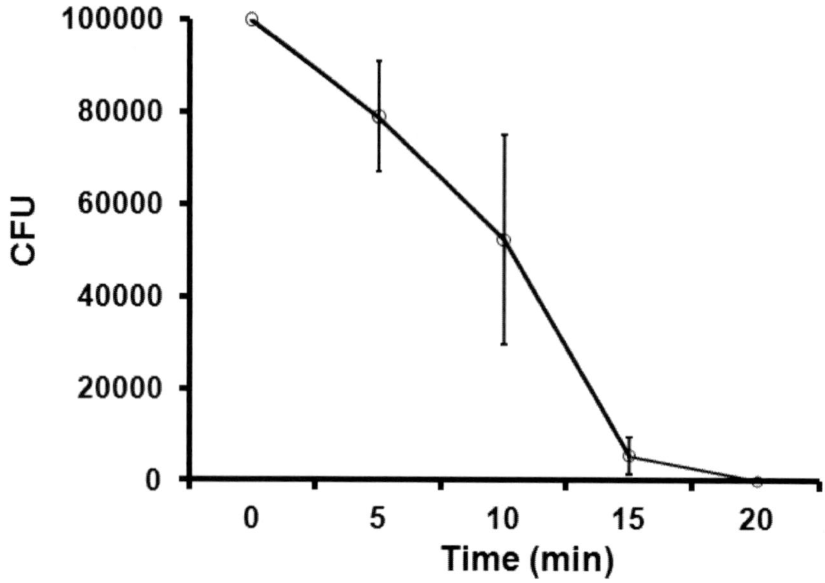

Figure 4. Effect of GM on *L. monocytogenes*. *L. monocytogenes* (1×10^5 CFU) are incubated with CM containing GM (2.5-fold MIC) and the CFU are determined every 5 min. Data represent CFU in each tube. Viable *L. monocytogenes* become undetectable 20 min after incubation.

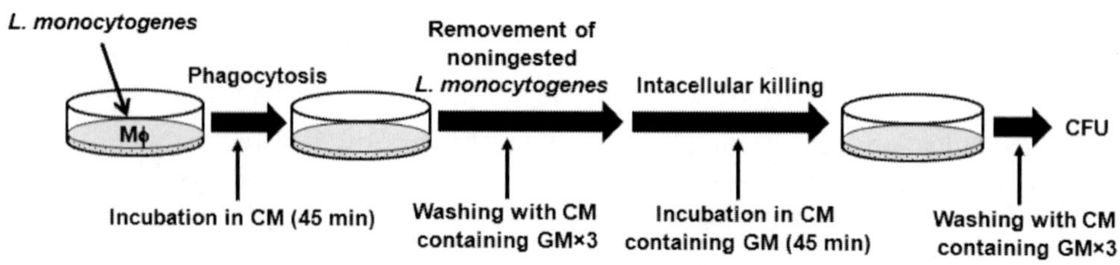

Figure 5. Experimental procedure for determination of optimal condition of GM. Mϕ are incubated with *L. monocytogenes* for 45 min. Subsequently, cells are washed 3 times with CM containing GM (2.5-fold MIC) to remove noningested *L. monocytogenes*. Cells are further incubated in CM containing GM (2.5-fold MIC) for 45 min, treated with 0.5% saponin, plated on agar plates after sonication, and the CFU are determined.

The optimal incubation period should also be determined by counting CFU in Mϕ at different time points. In our laboratory, it appears that the number of viable *L. monocytogenes* in Mϕ is the highest after 45 min incubation and the number of viable *L. monocytogenes* in Mϕ is markedly reduced after 90 min incubation (Kaneko, unpublished observation). Therefore, incubations for 45 min and 90 min are recommended for determination of phagocytic and listericidal activities of Mϕ, respectively.

Conclusion

This review describes several methods for determination of phagocytic and listericidal activities of Mϕ. Although a couple of methods recently employed are probably reproducible, expensive materials, expensive equipment and complicated methods are required. Moreover, it is impossible to determine both activities simultaneously. In contrast, in CFU-counting method expensive materials and expensive equipment are not required, and both phagocytic and listericidal activities of Mϕ can simultaneously be determined. Although it has been reported that GM kills not only extracellular, but also intracellular, bacteria, optimal condition of GM kills only extracellular bacteria. Accordingly, we assume that CFU-counting method should be reevaluated.

REFERENCES

[1] Cossart, P. (2011). Illuminating the landscape of host–pathogen interactions with the bacterium *Listeria monocytogenes*. *Proc. Natl. Acad. Sci. USA,* 108, 19484-19491.

[2] Mead, P. S.; Slutsker, L.; Dietz, V.; McCaig, L. F.; Bresee, J. S.; Shapiro, C.; Griffin, P. M.; Tauxe, R. V. (1999). Food-related illness and death in the United States. *Emerg. Infect. Dis.*, 5, 607-625.

[3] Adak, G. K.; Long, S. M.; O'Brien S. J. (2002). Trends in indigenous food-borne disease and deaths, England and Wales: 1992 to 2000. *Gut,* 51, 832-841.

[4] Harty, J. T.; Schreiber, R. D.; Bevan, M. J. (1992). CD8 T cells can protect against an intracellular bacterium in an interferon γ-independent fashion. *Proc. Natl. Acad. Sci. USA,* 89, 11612-11616.

[5] D'Orazio, S. E.; Troese, M. J.; Starnbach, M. N. (2006). Cytosolic localization of *Listeria monocytogenes* triggers an early IFN-γ response by $CD8^+$ T cells that correlates with innate resistance to infection. *J. Immunol.*, 177, 7146-7154.

[6] North, R. J.; Conlan, J. W. (1998). Immunity to *Listeria monocytogenes*. *Chem. Immunol.*, 70, 1-20.

[7] Pinto, A. J.; Stewart, D.; van Rooijen, N.; Morahan, P. S. (1991). Selective depletion of liver and splenic macrophages using liposomes encapsulating the drug dichloromethylene diphosphonate: effects on antimicrobial resistance. *J. Leukoc. Biol.*, 49, 579-586.

[8] Conlan, J. W.; North, R. J. (1994). Neutrophils are essential for early anti-*Listeria* defense in the liver, but not in the spleen or peritoneal cavity, as revealed by a granulocyte-depleting monoclonal antibody. *J. Exp. Med.*, 179, 259-268.

[9] Czuprynski, C. J.; Brown, J. F.; Maroushek, N.; Wagner, R. D.; Steinberg, H. (1994). Administration of antigranulocyte mAb RB6-8C5 impairs the resistance of mice to *Listeria monocytogenes* infection. *J. Immunol.*, 152, 1836-1846.

[10] Conlan, J. W. (1997). Critical roles of neutrophils in host defense against experimental systemic infections of mice by *Listeria monocytogenes*, *Salmonella typhimurium*, and *Yersinia enterocolitica*. *Infect. Immun.*, 65, 630-635.

[11] Garin, J.; Diez, R.; Kieffer, S.; Dermine, J. F.; Duclos, S.; Gagnon, E.; Sadoul, R.; Rondeau, C.; Desjardins, M. (2001). The phagosome proteome: insight into phagosome functions. *J. Cell Biol.,* 152, 165-180.

[12] Fang, F. C. (2004). Antimicrobial reactive oxygen and nitrogen species: concepts and controversies. *Nat. Rev. Microb.*, 2, 820-832.

[13] Ray, K.; Marteyn, B.; Sansonetti, P. J.; Tang, C. M. (2009). Life on the inside: the intracellular lifestyle of cytosolic bacteria. *Nat. Rev. Microbiol.*, 7, 333-340.

[14] Cossart, P.; Vicente, M. F.; Mengaud, J.; Baquero, F.; Perez-Diaz, J. C.; Berche, P. (1989). Listeriolysin O is essential for virulence of *Listeria monocytogenes*: direct evidence obtained by gene complementation. *Infect. Immun.*, 57, 3629-3636.

[15] Nishibori, T.; Cooray, K.; Xiong, H.; Kawamura, I.; Fujita, M.; Mitsuyama, M. (1995). Correlation between the presence of virulence-associated genes as determined by PCR and actual virulence to mice in various strains of *Listeria* spp. *Microbiol. Immunol.*, 39, 343-349.

[16] Harty, J. T.; Bevan, M. J. (1995). Specific immunity to *Listeria monocytogenes* in the absence of IFN-γ. *Immunity,* 3, 109-117.

[17] Huang, S.; Hendriks, W.; Althage, A.; Hemmi, S.; Bluethmann, H.; Kamijo, R.; Vilcek, J.; Zinkernagel, R. M.; Aguet, M. (1993). Immune response in mice that lack the interferon-γ receptor. *Science,* 259, 1693-1694.

[18] Drevets, D. A.; Campbell, P. A. (1991). Roles of complement and complement receptor type 3 in phagocytosis of *Listeria monocytogenes* by inflammatory mouse peritoneal macrophages. *Infect. Immun.*, 59, 2645-2652.

[19] Drevets, D. A.; Canono, B. P.; Campbell, P. A. (1992). Listericidal and nonlistericidal mouse macrophages differ in complement receptor type 3-mediated phagocytosis of *L. monocytogenes* and in preventing escape of the bacteria into the cytoplasm. *J. Leukoc. Biol.*, 52, 70-79.

[20] Utermöhlen, O.; Karow, U.; Löhler, J.; Krönke, M. (2003). Severe impairment in early host defense against *Listeria monocytogenes* in mice deficient in acid sphingomyelinase. *J. Immunol.*, 170, 2621-2628.

[21] Peck, R. (1985). A one-plate assay for macrophage bactericidal activity. *J. Immunol. Meth.*, 82, 131-140.

[22] Espaze, E. P.; Reynaud, A. E. (1988). Antibiotic susceptibilities of *Listeria*: *in vitro* studies. *Infection,* 16, 160-164.

[23] Bryskier, A. (2005). Antimicrobial agents: antibacterials and antifungals. Washington, DC: ASM Press.

[24] Vakulenko, S. B.; Mobashery, S. (2003). Versatility of aminoglycosides and prospects for their future. *Clin. Microbiol. Rev.*, 16, 430-450.

[25] Portnoy, D. A.; Suzanne, J. P.; Hinrichs, D. J. (1988). Role of hemolysin for the intracellular growth of *Listeria monocytogenes*. *J. Exp. Med.,* 167, 1459-1471.

[26] Singh, R.; Jamieson, A.; Cresswell, P. (2008). GILT is a critical host factor for *Listeria monocytogenes* infection. *Nature,* 455, 1244-1247.

[27] Emoto, M.; Yoshida, T.; Fukuda, T.; Kawamura, I.; Mitsuyama, M.; Kita, E.; Hurwitz, R.; Kaufmann, S. H. E.; Emoto, Y. (2010). α-galactosylceramide promotes killing of *Listeria monocytogenes* within the macrophage phagosome through invariant NKT-cell activation. *Infect. Immun.,* 78, 2667-2676.

[28] Huynh, L.; Wang, L.; Shi, C.; Park-Min, K. H.; Ivashkiv, L. B. (2012). ITAM-coupled receptors inhibit IFNAR signaling and alter macrophage responses to TLR4 and *Listeria monocytogenes*. *J. Immunol.*, 188, 3447-3457.

[29] Davis, M. J.; Gregorka, B.; Gestwicki, J. E.; Swanson, J. A. (2012). Inducible renitence limits *Listeria monocytogenes* escape from vacuoles in macrophages. *J. Immunol.*, 189, 4488-4495.

[30] Drevets, D. A.; Canono, B. P.; Leenen, P. J.; Campbell, P. A. (1994). Gentamicin kills intracellular *Listeria monocytogenes*. *Infect. Immun.*, 62, 2222-2228.

[31] Ohya, S.; Xiong, H.; Tanabe, Y.; Arakawa, M.; Mitsuyama, M. (1998). Killing mechanism of *Listeria monocytogenes* in activated macrophages as determined by an improved assay system. *J. Med. Microbiol.*, 47, 211-215.

[32] Kaneko, M.; Emoto, Y.; Emoto, M. (2014). Acquired resistance of *Listeria monocytogenes* in and escaped from liver parenchymal cells to gentamicin is caused by being coated with their plasma membrane. *Microbes Infect.*, 16, 237-243.

[33] Francis, G.; Kerem, Z.; Makkarand, H. P.; Becker, K. (2002). The biological action of saponins in animal systems: a review. *Br. J. Nutr.*, 88, 587-605.

INDEX

C

D

E

F

G

H

I

J

K

L

M

N

O

P

Q

T

U

V

W

Y